Springer Climate

Series Editor

John Dodson, Institute of Earth Environment, Chinese Academy of Sciences, Xian, Shaanxi, China

Springer Climate is an interdisciplinary book series dedicated to climate research. This includes climatology, climate change impacts, climate change management, climate change policy, regional climate studies, climate monitoring and modeling, palaeoclimatology etc. The series publishes high quality research for scientists, researchers, students and policy makers. An author/editor questionnaire, instructions for authors and a book proposal form can be obtained from the Publishing Editor.

Now indexed in Scopus® !

Marko Hakovirta

Carbon Neutrality

Follow the Money

Springer

Marko Hakovirta
College of Natural Resources
North Carolina State University
Raleigh, NC, USA

ISSN 2352-0698 ISSN 2352-0701 (electronic)
Springer Climate
ISBN 978-3-031-45201-7 ISBN 978-3-031-45202-4 (eBook)
https://doi.org/10.1007/978-3-031-45202-4

This Springer imprint is published by the registered company Springer Nature Switzerland AG
The registered company address is: Gewerbestrasse 11, 6330 Cham, Switzerland

Paper in this product is recyclable.

To my wife and sons, this book is dedicated to you, you have been my unwavering support, my inspiration, and my motivation throughout my career.
To my wife, your understanding and belief in me have been the bedrock of my accomplishments. Your faith in my abilities and your encouragement have moved me forward, even though you were diagnosed to have a critical illness during this project. Thank you for being by my side and for being my friend and partner in both life and my career.
To my sons, you are my greatest joy and my proudest accomplishment. Your humor, boundless energy, and inquisitive minds remind me daily of what truly matters. Together, you have been my source of happiness and inspiration.
To my mother and father, who both have faced declining health during this book project. You both have been my pillars of strength and the driving force behind my endeavors.
With all my love,

Marko Hakovirta

Acknowledgments

I would like to express my sincere gratitude to all those who have contributed to the creation and completion of this book. Their support, expertise, and dedication have been invaluable throughout the entire process.

First and foremost, I would like to thank the Technical Research Centre of Finland (VTT) for their invitation sponsorship to join VTT as a visiting professor and my dear colleague and host during my sabbatical; Prof. Jussi Manninen. Your continuous support to this book and dedication to the field always inspires me and truly made this book possible. Thank you to all researchers and leaders in various institutions and corporations including, for example, VTT, Business Finland, Implement Consulting Group, Metso, Valmet, Stora Enso, UPM-Kymmene, Sitra, Stanford University, University of Turku, University of Tampere, University of Helsinki, and North Carolina State University. The people I have had the privilege to discuss and interview have shared their knowledge and insights, that have shaped the content of this book. Their expertise and commitment to the topic have been instrumental in providing comprehensive and accurate information.

Special thanks goes to Hanna Riehl and Laura Virtanen who helped me in data collection and analysis in terms of global investments in carbon neutrality. Also, I want to thank Krista Kovanen for her help in entrepreneurship research and my students Ann Winstead and Bryanna Lawry who supported me in surveying and analyzing corporate policies and targets.

I would also like to extend my gratitude to the reviewers and editors who have provided valuable feedback and suggestions, helping to refine and enhance the quality of the content.

I am also grateful to my colleagues, friends, and family members who have provided encouragement, support, and understanding throughout the writing process. Special thanks goes to Dr. Markku Karlsson, my career long mentor. Your unwavering support and belief in this book project have been a constant source of motivation.

Finally, I would like to express my appreciation to the readers of this book. Your interest and engagement in the topic of climate change and mitigation efforts are crucial in driving positive change and shaping a more sustainable future.

Acknowledgments

Contents

About the Author

Dr. Marko Hakovirta's research is focused on sustainable business, responsible innovation, and entrepreneurship with an emphasis on the bioeconomy. He also publishes research in interdisciplinary work on topics of processing and engineering of renewable biomaterials and valorization of industrial and municipal waste. He teaches sustainability and innovation management and life cycle analysis in undergraduate and graduate levels.

Dr. Hakovirta brings his more than 20 years of leadership experience working on local, national, and international topics in research, education, and strategic sustainability and innovation management. He engages with global businesses with an innovative and proactive approach, aligning sustainability and value creation.

Dr. Hakovirta is Fulbright specialist, TAPPI Fellow and EVA Business Fellow and is former Board Member of TAPPI and Society of Wood Science and Technology. He also advises and consults small and large corporations in bioeconomy.

Dr. Marko Hakovirta has previously worked as Senior Vice President, Innovation and R&D at Stora Enso Corporation and he served as Group Vice President, Technology, Environment, and Quality at Metso Corporation. He has also held leadership positions at Auburn University and Georgia Tech and has held fellowships at European Particle Physics Laboratory (CERN) and at Los Alamos National Laboratory. Dr. Hakovirta received his Ph.D., M.Sc., and B.Sc. degrees from the University of Helsinki, Finland, and an MBA from Emory University, Atlanta.

Chapter 1
Introduction Brief History of Carbon Neutrality

Abstract In this chapter, we explore the concept of carbon neutrality, also known as net zero carbon emissions, which involves balancing the amount of CO_2 emissions released into the atmosphere with an equivalent amount of CO_2 removal or offsetting. We also delve into the fragility of Earth and its history of facing various climate issues that have posed risks to life. Over millions of years, the Earth's climate has continuously changed due to natural processes, such as volcanic activity, changes in the Earth's orbit, solar radiation variations, and geological events. However, in recent times, human activities have emerged as a significant driver of climate change, leading to unprecedented shifts in the Earth's climate at a faster pace. Throughout this chapter, we examine the complex interplay between climate change and human history, exploring the impact of human activities on the climate and the importance of achieving carbon neutrality to address the pressing climate issues. The lessons from history and the challenges and opportunities in addressing climate change further emphasize the need for collective action and sustainable practices to safeguard our planet and its ecosystems.

Keywords Birth of civilization · Evolution of life · Rise of industrialization · Global environmental crises and climate change · Anthropogenic climate change

1.1 Early History

Our planet has existed in its current form for about 4.6 billion years. The planet Earth was formed by matter, composed of dust and gas, and left over from the birth of our sun. The same process of coalescing from the enormous disk-shaped solar nebula created all the planets and moons in our solar system. We literally are of stardust, and as such looking at the whole process of planetary formation, it is a marvel by any definition that Earth and our solar system exist.

M. Hakovirta, *Carbon Neutrality*, Springer Climate,
https://doi.org/10.1007/978-3-031-45202-4_1

The dawn of life on Earth is even bigger miracle. Based on the evidence of oldest fossilized microorganisms, life already existed some 3.5 billion years ago. Some evidence even exists that the most primitive forms of life existed a few hundred million years prior in certain submarine-hydrothermal vents. Regardless of the details, we all can agree that the foundations of life are fragile, and unless science had not rigorously been looking for answers to our mere existence, we would call all this almost humorous deception.

The evolution of life took its time and started from a microbial mat of multilayered microorganisms such as bacteria and archaea. Both are only a few micrometers in length and the latter is able to live in extreme environments including high levels of temperature, pH, radiation, and salinity. When these "mats of life" were formed, there needed to be a distinct equilibrium of disruptive physical dynamics and the microbial responses to enable them to survive. The next major step in the evolution was the creation of photosynthesis process in which energy from sunlight is transformed to chemical energy that enables to fuel the living organism to have different activities. It is estimated that the timing for all this is about 3.2 billion years ago. Some of the chemical energy produced in photosynthesis is in carbohydrate molecules such as sugars and starch. They are both prerequisites for more advanced life and products of CO_2 and water which the photosynthesis process "puts together." This process is the main activity of most plants and algae on Earth and was responsible for the *Great Oxygenation Event* (*GOE*) that took place after about a billion years. During the Great Oxygenation, the Earth's atmosphere changed from practically oxygen-free state to abundance of oxygen that was absorbed in oceans, seabed rock, land surfaces and accumulated in the atmosphere. Interestingly, this critical to live event was due to cyanobacteria that produced oxygen and created enough chemical energy storage to later develop multicellular life form. We seem to be in high gratitude to *cyanobacteria*; however, the oxygenation was not good for all living microorganisms as the high oxygen concentration in the biosphere—the parts where life exists on Earth—caused the extinction of many anaerobic microorganisms.

The next milestone in evolution was *eukaryotes* that are organisms with cells that have nuclei and that have subunit structures with specific functions. This happened some 2 billion years ago, and specialization of the living organisms continued with the large quantities of oxygen present. The next steps included the development of sexual reproduction that most complex macroscopic organisms use. And about a billion years ago land-based plants started developing from the primitive algal mats through the evolution steps from green algae to non-vascular land plants and vascular plants with spores and seeds all the way to gymnosperms that include the conifers and finally the flowering plants. During this same period, around 700 million years ago, clusters of cooperating cells that were also specialized to different tasks started evolving into the first animals. Around 540 million years ago life became more dynamic and countless new species emerged and started filling the oceans and land. Interestingly, the story of land plants was quite dramatic as their successful growth impacted the biosphere so much that it is believed it was one of the reasons that created a so-called *Late Devonian extinction* event. This happened about 365 million years ago when even 70% of all living species in oceans and land became extinct possibly due to transforming soil into organic-rich sediments and with their run-offs algal blooms that transformed the oceans to be anoxic or oxygen-deprived thus creating major loss of marine life. Some theories also suggest that there was major reduction of CO_2 levels due to high plant and algae activity, and therefore, it created global cooling and reduced sea levels.

The life on Earth evolved with five distinct extinction events between 450 and 66 million years ago. The first one is called *Ordovician–Silurian extinction event* and was according to the latest research due to global warming, related to volcanism, and low oxygen conditions. The second was the *Late Devonian extinction*, with several causing events in series as discussed earlier. The third was *Permian–Triassic extinction event*, which also was the largest extinction even on Earth killing more than 90% of all species. It took place about 252 million years ago, and current scientific consensus explains the event to be caused by elevated temperatures due to large quantities of CO_2 in the atmosphere. The reason for this dramatic increase would have been due to massive volcanic eruptive events in Siberian region with volcanic rock (Siberian Traps). The fourth event was *Triassic–Jurassic extinction event* which happened 201 million years ago. During this period, most non-dinosaurian animal species were extinct (70–75% of all species) giving dinosaurs excellent non-competitive living environment. The last period of extinction was *Cretaceous–Paleogene extinction event*. More than 60 million years ago all dinosaurs that were not avian (could not fly) were extinct, and after this event, the now dominant group of mammals that descended from the vertebrate animals and mammals like reptiles and birds originated from surviving hollow-boned and three-toed limbed dinosaurs. The reason for this latest event is generally agreed to have been caused by a massive comet or asteroid with a size of 10 to 15 km in width: the stellar object possibly hitting the area of Yucatán Peninsula in the Gulf of Mexico. The impact produced atmospheric changes that caused major effect to the climate and acidified the oceans. The caused long-term effects produced ecological disruption leading to extinction. Another event during the same time period could have been also a contributing factor. In west-central India, so-called *Deccan traps* went through major volcanic eruptions and the release of volcanic gases such as sulfur dioxide and carbon monoxide contributed to the climate change as well. During this period, it has been estimated that the global average temperature dropped 2 °C.

Eventually, the long 6-million-year process of human evolution started from our ape-like ancestors. Our skills and capabilities evolved gradually, and we were able to walk on two feet around 4 million years ago. The advancement took its time, and finally, more complex abilities such as use of tools and speech and art started around 100,000 years ago. The early humans had many ancestral species during the evolutionary progress and eventually died out. The original African location for early humans was not always ideal due to dramatic climate fluctuations, changes in landscape and food supply. Although early upright ape *Homo erectus* started migration already around 2 million years ago, the *Homo sapiens* migrated from Africa by crossing across the Middle East and by spreading throughout Asia, Oceania, and Europe. All this took place around 60,000–40,000 years ago.

1.2 Birth of Civilization

The human civilization progressed through three different stages: the *preagricultural* (hunting and gathering), the *agricultural,* and the *industrial* stages. The first ever human civilization was established in Mesopotamia. The location of this *"Cradle of Civilization"* is specifically region of Western Asia and within the Tigris–Euphrates River system which discharges into the Persian Gulf. The name Mesopotamia means in Greek literally in between rivers but the range of its influence expanded much beyond it. This region at large (beginning of civilization) is currently defined by Israel, Iraq, Jordan, Lebanon, parts of Iran, Syria, and Turkey and was originally called *"Fertile Crescent."* It was first populated about 10,000 Before Common Era (BCE) because the fertile river system created favorable conditions that supported biodiversity and expansion of population and the development of agriculture. In a few thousand years by around 5000 BCE, the agriculture and the domesticating farm animals were already widespread and irrigation systems for agricultural crops were in place and fully developed. These thousands of years of evolution of civilization contributed to the world culture including the development of knowledge such as writing, commerce, and science. Many important engineering inventions were also discovered such as the invention of wheel.

The other hot spot in the *"Fertile Crescent"* for the early development of civilization was ancient Egypt. The development of larger inhabitation that was dependent on agriculture started already some 7000 years ago and the evolution of the ancient Egyptian civilization peaked during a 3000-year era between 3300 and 330 BCE as a preeminent civilization of its time. Many of the major achievements this period produced include hieroglyphic writing, mathematical system for predicting cyclical natural events, developments in medicine, highly advanced construction and quarrying techniques, and several agricultural technologies including irrigation.

Further away in the east, the *Indus Valley civilization* was born and was the earliest known Bronze Age society in the South Asia. The accomplishments of this so-called Harappan civilization ranged from 3300 to 1300 BCE with a peak of maturity in around 2600–1900 BCE. Again, the success of this civilization was founded on the river system and its biodiverse ecosystem. The area was ideal for growing wheat, barley, and oats which enabled the development of a sophisticated agricultural system. The cities developed in the region were highly developed and had substantial architectural progress including indoor plumbing and drainage systems. And the society by enlarge was responsible for developing astrology, mathematics, and the development of a measuring system for weight.

Another "Cradle of Civilization" in the Old World (the world known to its inhabitants before the discovery of the Americas) was ancient in China. The origins of inhabitation in China can be traced back to 1.7 million-300,000 years ago. Archeological evidence has shown that the early populations had the knowledge to use stone tools and utilize fire. A generally accepted fact is that the Chinese "Cradle of Civilization" is situated in the *Yellow River Valley,* and evidence of agricultural development and villages can be found from as far as 5000 BCE. The evolution of from farming communities to centralized government started from the foundation of the Xia Dynasty, which existed between 2070 and 1600 BC. This was also a foundation of a hereditary system, with succession of different dynasties all the way to modern times and ending with Qing Dynasty in 1912. These highly developed societies contributed globally to the various aspects of knowledge and technology. Most known great inventions that were developed in China and have had a global impact to modern world include papermaking, printing (nearly 600 years before Johan Gutenberg's invention in Europe), gunpowder, and the compass.

Ancient Americas is not by definition included to the birth of Old World civilization but clearly needs to be discussed as part of the development of human civilization. The migration of early humans to and across the Americas (Paleo-Indian migration) started from the *Bering land bridge* between Siberia and Alaska somewhere between 40,000 and 17,000 years ago. The science explaining the migration is still disputed but the general agreement is that the migration happened by both East Asian and Western Eurasians population. After a several thousand years, the migration had advanced to most of the Americas, and the foundation of early civilization of Americas began somewhere around 6000 BCE. By that time, the domestication of maize and the development of mixed hunter-gatherer and fishing culture had been developed. Around 3000 BCE–1800 BCE, more complex civilization emerged in the northern coastal region of Peru (Norte Chico). This earliest Ancient Americas civilization already had developed fishing nets and textiles cotton and had a distinct government structure. After the Peru's Norte Chico civilization, many other centralized civilizations started emerging. The most prominent and historically important was the Mesoamerica area that covers most of the southern North America and most of Central America.

The first Mesoamerican civilization was *Olmec society* that emerged in 1500 BCE, and it catalyzed several other civilizations including *Mayan* (emerging around 700 BCE), *Aztecs* (1300–1500 CE, central Mexico) and *Incas* (1400–1600 CE, Peru region). The societies of the Mesoamerica were highly developed with cities that were of similar size as the Old World's counterparts with even 35,000 inhabitants in the Aztec capital, Tenochtitlan. Many agricultural developments took place, including breeding of maize, and domestication of agricultural products such as tomatoes, potatoes, beans, avocados, and cocoa. Eventually, many of these advancements migrated to north (Mississippian culture, 800 CE–1600 CE) and with the conquistadors and European settlers to the Old World. Ancient Americas was in a critical role in the societal transformation towards urban cultures and the integration of Indigenous, European, African, and Asian cultures and civilization that the New World is founded on.

1.3 Rise of Industrialization and Global Economy

The major development step in the human civilization was the rise of *Industrial Revolution* that transformed economies founded on agriculture and handicrafts. It started as the development and transition to an era of manufacturing and took place primarily in Europe and the USA, within a period between 1760 and 1840. The root causes for this rapid development period included the advancement of the agriculture in the eighteenth century that produced excess of food driving the price down and creating disposable income in England for purchasing manufactured goods. The colonial system of the British Empire also gave an access to various raw materials used in manufacturing textiles and various other goods.

Other factors for industrial growth in Britain included large middle class, mobile workforce, the river system that enabled an effective supply chain, and continuous technological innovation. The continuous and accelerated development of manufacturing processes, machines, and tools increased the productivity of agriculture, textiles, and transportation. This new industrialized world enabled the citizens to gain higher food and nutritional consumption and enabled increased population growth and life expectancy. The Industrial Revolution demanded more energy, which was covered by the transitioning to coal-based energy and leading also to the invention of high-pressure steam engine. The steam engines enabled to pump water from mines and even further increase the access to low-cost coal. The use of coal was additionally beneficial in the advancements in metallurgy. This led to the creation of wrought iron for machinery and many other industry applications. As the industrialization progressed, the world started becoming progressively dependent on coal, oil, and natural gas. This eventually led to the advancement of climate change caused by human activity also called *anthropogenic climate change.*

Many of these industrialization-related developments occurred during a fairly short time period, and the speed of change became exponential. The progress of interconnected innovations diffused to most of the Europe and North America and created a tremendous economic growth. The industrial globalization started in the mid-1800s and was enabled by global labor mobility and efficiency and low cost of the newly developed supply chains. The large-scale cross-border production of goods grew in combination with technology and organizational advancements and created a competitive advantage to the Western companies, making it possible for them to penetrate global markets. This led to outsourcing of manufacturing and labor and benefiting on the adventitious cost-effectiveness operating in developing nations. These global enterprises drove down cost of production. Centers of inexpensive labor and production sites were established to maximize profits. As global economy and industrialization evolved, this approach allowed the developing nations to also learn and develop their own manufacturing processes and related technologies, in addition to the global business skills. The shift of the global economic power to the most populous developing nations eventually started due to rapid rise in productivity and the development of local policies on competition, accumulation of capital, domestic savings, and investments in infrastructure. Some of the developing nations were not able to seize the opportunities, and therefore, only selected economies grew faster and more consistently than others. The winners among the developing nations witnessed remarkable growth and were able to develop their industrial sectors as global leaders in manufacturing and thus became a major source for anthropogenic climate change.

This transformation of global economic landscape was not only affecting the way work was organized globally or the way goods were manufactured. One of the large societal impacts was also on how people interacted and communicated with each other globally. It changed the global cultural arena, the human connection to the environment and geopolitical landscape. The Industrial Revolution was not only beneficial to human civilization but also had its negative impacts. The most powerful grouping of the countries or emerging economies that have been most successful in the development of global economy include Brazil, Russia, India, and China (also called BRIC countries). Their growth has been remarkable and in 2020 their Gross Domestic Product (GDP) amounted to nearly 20.3 trillion US dollars. As a reference point, the US GDP in 2020 was $20.9 trillion. With the increased wealth, access to low-cost food, and the affordability and advancement of medicine and health care, there has been a major upturn in the world population numbers. In 1750, the beginning of Industrial Revolution, the world population totaled 791 million. In 1850, it was already 1.26 billion, and by the end of the twentieth century, it had reached 6.1 billion. The growth is exponential and regional and projected to continue to reach 9.7 billion in 2050. The regionality is also distinct; in 2016, Asia's population was 4.4 billion and is anticipated to reach 5 billion people by 2050. However, the European population was 510 million in 2016 and is estimated to stay at the same level in 2050. Other major changes in the world population include aging and especially in the developed countries of the West. Looking at this in a global level, in 1990, there were 0.5 billion people aged over 60 years old, in 2017, there were 1 billion, and by 2050, the same number will be 2.1 billion. It can be stated that the population growth in combination with aging is causing shift of the economic power, which will have an effect in global business and investment trends and the society.

Based on the GDP, the emerging economies are now the growth markets, similarly as Britain, Europe, and North America were 200 years ago. The change in the roles in global economy has taken place rapidly. The developing countries have transitioned from manufacturers of goods supplied to developed countries, to economies with growing middle class having purchasing power that makes them attractive market for consumer goods and services. This market has gained growth fast and has increased twofold since the 1990s. The emerging economies' share of the global growth is currently 80%.

Although China is still considered as a developing country, it has become the global manufacturing hub and is already the second-largest global economy measured by GDP. Amazingly only 20 years ago, China's GDP was 10% of the US GDP. The current Chinese population number is 1.4 billion, and the country is going through major urbanization phase. China will have over 200 cities with over one million inhabitants by 2030. As a comparison, the USA has at present ten cities of that size. The economic growth of China has recently slowed down with debt levels (private and government) rising dramatically. Currently, the total debt to GDP ratio is higher than the USA, and the private debt has reached the level of 200% of the GDP. Regardless of some challenges, Chinese business growth is impressive and includes 100 companies with $1 billion valuation (also called unicorns). The fastest-growing industries in China are medical device manufacturing, transportation industry, e-commerce, hotel and tourism industry, Internet services, and online gaming. The largest at present are still quite manufacturing focused and include copper ore mining, building and construction, real estate development, online shopping, infrastructure construction (bridge, tunnel, subway), software and steel industry.

The *economic power shift* relates strongly to sustainability and specifically carbon neutrality. As the developed nations have gone through the journey of development of manufacturing technologies, and global markets, they also have been able to afford to put in place strict environmental regulations and policies to achieve sustainability goals. These policies have had a major impact in the global value chains, affecting the design and production of various products and services, and their trade. The policies and regulations are however not global, and in many of the developing countries, the focus is more on building strong economies. The steps towards stricter environmental policies and regulations are slow and continue to develop asymmetric actions to mitigate environmental sustainability and climate change. The global corporate responsibility actions, non-governmental organizations (NGOs) activities and increasing consumer awareness and interest in sustainability have not been able to create enough momentum to have realistic transition to a more carbon–neutral or low-carbon society that is needed to effectively and timely address the current and projected climate change impacts. A more resource-efficient and biodiverse society is still a far too long away goal.

From a global perspective, *carbon neutrality* can be an opportunity to countries (developed or developing) that have already embraced sustainability and continued to invest in it. Environmental technology exports, lower future carbon offset costs (where regulated), or reduction in greenhouse gases (GHG) will be beneficial for those countries' employment and growth. The world of finance and banking is extremely important and impactful, and the current sustainability trends are increasing the need for companies to use Environmental, Social and Governance (ESG) criteria to secure capital investments. The overall increase in ESG and socially conscious investing will also limit growth in several industrial sectors and also in geographical areas in which sustainability is not progressing as expected. The inability to adapt to global warming and climate change mitigation by not having ambitious but also realistic targets or implementation roadmaps can create a reversal effect to the recent decades' advances in wealth and prosperity. These effects may not necessarily be related to business growth and economic issues but rather our planet's reactions to global warming through rise of sea level, droughts, and severe weather phenomena that do not differentiate the location based on GDP and can therefore create tremendous stress to the most vulnerable societies and their infrastructure.

1.4 Global Environmental Crises and Climate Change

The current global environmental crises have spread rapidly and are centered around climate change and scarcity of natural resources. During the past 100 years, the *anthropogenic climate change* has intensified the global average temperature levels already by 1 °C (or 1.8 °F), and the increase in sea level has been about 0.2 m (or 8 inches). In the recent past, the world has experienced some of the highest recorded global average temperatures. The rising temperatures increased the extreme weather phenomenon with record frequency and intensity of North Atlantic hurricanes and typhoons in the north-west Pacific. Other impacts include increase in forest fires and in some areas devastating impact to agriculture. The melting ice in Greenland and North Pole has caused flooding and created challenged to fishing and farming in coastal areas. The stress has therefore been most severe to rural and vulnerable poorer communities and therefore enhancing the migration to high population dense urban areas. The global warming also creates catastrophic events that can speed up the process of warming. For instance, soils in the permafrost region are estimated to hold close to two time the carbon the atmosphere does. With a total of almost 1600 billion tons of carbon stored, when released the amount can by itself make the climate change unstoppable.

The advancement of our civilizations from the humble beginnings of humankind has created the Industrial Revolution that opened a Pandoras box of dependence on using fossil fuels, leading to the destructive greenhouse gases release and global warming. The greenhouse gases from human activity have enlarged dramatically since the time of the Industrial Revolution. Based on various predictions and climate models, these emissions continue the same trajectory regardless of the governmental and private sector efforts. Looking at the current governmental policies, and their implementation progress, the average temperatures will rise by an estimated 3.9 °C by 2100 in comparison with the preindustrial levels (1850–1900). This will make considerable and permanent damage to the global ecosystem and natural resources.

In this book, we will look more closely at the big picture on how the global policies, regulations, private sector targets, and ESG finance are driving carbon neutrality and climate change mitigation. We will discuss the solutions we have and the emerging opportunities that the governments and private sector have in order to reach the challenging targets we are facing. We try to find answers to different scenarios regarding whether we just have to adapt or we actually able to change the direction of human activity-based greenhouse emissions. Some of the solutions are technological and some in novel business models such as circular economy or both. The target for this book is to give readers a holistic understanding on the tools we have and where we stand. The rest is up to us and our children.

Chapter 2
Impetus for Carbon Neutrality–Frames of Reference

Abstract Climate change presents a significant global challenge primarily caused by the rise of greenhouse gases in the Earth's atmosphere. These gases act as a natural insulating layer, vital for supporting life on our planet. The carbon cycle normally regulates CO_2 levels, but human caused, or anthropogenic emissions disrupt this balance, leading to global warming. The value chains of industries that are serving societal needs contribute to emissions, necessitating a comprehensive sustainability approach. The motivation to address climate change stems from the growing awareness of its severe impacts on the environment, ecosystems, and human societies. These impacts include more frequent and intense extreme weather events, rising sea levels, disruptions in agriculture and water resources, threats to biodiversity, and increased health risks. In this chapter, we explore various aspects related to anthropogenic emissions and discuss concepts for enhancing the value chain to mitigate climate change.

Keywords Global climate change · Greenhouse effect · Slow carbon cycle · Fast carbon cycle · Anthropogenic GHG emissions · Global warming potential · Value chain

2.1 Background

Climate change has been shaping planet Earth since the beginning of time. Extinction events (or biotic crises) have been caused by rapid decrease in earth's biodiversity due to climate and weather changes and have been part of life throughout its history. These life disruptive events have also created new evolutionary directions that have led to the origin of humanoids and homo sapiens and our dominant role on Earth. Climate change events have emerged in many forms, and they can be defined as long-term shifts and changes in weather patterns and extreme weather that are characteristic to local, regional, and global climate conditions. Looking at human civilization and

M. Hakovirta, *Carbon Neutrality*, Springer Climate,
https://doi.org/10.1007/978-3-031-45202-4_2

its history, the most important developments or even leaps in societal, technological, and economic conditions came in the best possible local or regional ecosystem environments.

Total civilizations have risen and vanished depending on these conditions. During the era of industrialization, with the help of technology, we have better adapter to drastic changes in climate and ecosystems. We have been able to tame nature and its course when it comes down to developing favorable local systems and microclimates for the beneficial development of societal and economic growth. Human made microclimates, that have been developed using artificial water reservoirs, damns, forest management, and geotechnical engineering have enables us to spread our footprint to regions where the conditions have traditionally been unfavorable. The list is long and displays both human ingeniousness, desire for expansion and greed.

As earlier described climate change has both a natural climate change component and an anthropogenic (human made) component. Clearly sometimes the distinction and definition may result critical thinking and discussion amongst the public and even the scientists. A scientific fact is that the atmospheric concentration of CO_2 has historically been quite stable at 250–280 parts per million (ppm) since the beginning of human civilization (10,000 years or so). Since the end of Industrial Revolution and the beginning of industrialization (1850s) the atmospheric CO_2 concentration has risen to the level of 412.5 ppm. The growth is exponential and highly alarming since only during the past 20 years it increased by 43.5 ppm (or 12%). The anthropogenic impact to global CO_2 levels is clear as these levels have risen about 100 times faster over the past 60 years compared to the last major natural climate change events that occurred when the previous ice age ended (11,000–17,000 years ago).

The science is able to estimate the historical CO_2 levels dating from tens of millions of years back by taking samples from cores of mud, deep from within the ocean floor. The microscopic fossils and ancient molecules that are confined and preserved in these "time capsules" conserve the history of CO_2 levels on our planet. Similarly, another technique that looks at the ice core samples and their CO_2 data over the last one million years are also showing comparable results. This well aligned information shows us that we currently have the same CO_2 level in our atmosphere as roughly 4 million years ago. Going further in the history the levels were for example at 1500 ppm level during the era of dinosaurs when there was no ice on poles and the average temperature was about 4–6 °C higher.

2.2 Global Climate Change and Greenhouse Gases

The history of Climate Science can be stated to date back to the time of discovering how to detect CO_2. The first detector was invented by Joseph Black, a medical student from Edinburgh. He realized that limewater can be used a CO_2 detector. This limewater instrument was further developed by Lord Cavendish and is better known as the Cavendish Apparatus. Eunice Newton Foote was an American scientist that in 1856 was able to proof the connection between atmospheric temperature and CO_2

concentration. Following her discovery, John Tyndall, a British physicist, researched the role of atmospheric gases in the greenhouse effect. His work helped to better understand that the three components: CO_2, ozone, and water vapor are critical for global temperature due to greenhouse effect. Swedish scientist Svante Arrhenius was able by 1890 to predict how the increased CO_2 concentration would increase the global temperatures and British Engineer, Guy Steward Callendar was able to show evidence that the global temperatures are rising, and in hindsight be able to predict quite accurately the temperature increase if CO_2 levels continue to rise.

Greenhouse effect was first studied in detail and described by mathematician Joseph Fourier. The effect is not only caused by CO_2, but other gases including methane (CH_4), nitrous oxide (N_2O), water vapor and several specific gases released by the industry: nitrogen trifluoride (NF_3), perfluorocarbons (PFCs), sulfur hexafluoride (SF_6) and hydrofluorocarbons (HFCs). As our planet gets continuously energy generated by the sun by means of a wide spectrum of electromagnetic radiation (averaged 340 W/m^2), some 30% of that energy is reflected back to space by atmospheric gases, water vapor (clouds), oceans, snow and ice and the Earth's land mass. The remaining energy (240 W/m^2) is absorbed by atmosphere, oceans and land but also is re-radiated back to space by a lower energy frequency (infrared radiation). This re-radiated frequency range happens to be highly effective for water vapor and greenhouse gases to absorb and thus creates a partial "heat wall" that keeps the lower atmosphere warm and the upper cooled. A small fraction of the radiated energy has a specific frequency to escape Earth (40 W/m^2). The rest of the energy goes through a complex layered process of absorption and re-radiation and finally creates the energy balance that is responsible for the average temperatures that support complex life on Earth. In actuality greenhouse gases we look at critically as the source of global warming are the reason for our planet to be able to sustain its delicate life.

2.3 Role of CO_2

Why is the CO_2 in such a critical role in greenhouse effect and Climate Change? Firstly, the CO_2 concentration is normally relatively low and continuous increase in concentration is now able to make a larger impact by closing off parts of the atmospheric window (frequency at which re-radiated energy is not re-absorbed and can escape to space). The resulting increased temperature then makes the atmosphere able to hold more gaseous phase of water (water vapor) and this will further increase the overall absorption of energy. In essence a thicker "wall" for greenhouse effect is created, further increasing an average temperature of our planet. Water vapor is self -regulating and with higher concentration and no additional heat, it will condense back to liquid by rain. The other greenhouse gases are however long lasting. In terms of CO_2 there are numerous natural processes taking it out from the atmosphere. It is estimated that 65–80% of CO_2 in the atmosphere is dissolved to the oceans in a few tens to hundreds of years. The remaining CO_2 then is absorbed by so called *Slow Carbon Cycle* over billions of years. The rest of the GHGs survive differently.

For example, methane is removed from the atmosphere by chemical reactions in 10 years. The nitrous oxide survives longer, and it will break down as it reaches the stratosphere and unfortunately also produces ozone-destroying reactions. In general, they survive in the atmosphere for about 100 years. The industrially made chlorinated or fluorinated gases can stay in the atmosphere for a range of a few years to a few thousand years.

We have now established the fact that CO_2 is a critical gas in Climate Change. It almost performs as a temperature controller for Earth. Let's look more closely at how this so-called *Carbon Cycle process* functions. This cycle is highly complex and includes biological, chemical, and physical factors. The CO_2 in our atmosphere is the form of carbon, the element that is the "foundation of life" on Earth. In addition to regulating our planets temperature it is a precursor for all life forming complex molecules including DNA. Our planet is a closed system and therefore all carbon moving in the cycle to and from the atmosphere is in a constant flow. Most of the carbon is situated on Earth's surface including sediment, soil, rocks and living organisms. In addition, oceans take in CO_2 by dissolving it or by processes that include photosynthesis.

The exchange of carbon between the Earth's atmosphere and the ocean occurs, with the ocean acting as a vital component of the global carbon cycle. It serves as a significant reservoir for CO_2, both in its surface waters and through long-term storage in its depths. The ocean's ability to capture and store carbon, along with coastal ecosystems like mangroves, marshes, and seagrasses, is referred to as *blue carbon*. The sources of release of carbon to the atmosphere include decomposition of organic matter, forest fires, volcanic activity, and the use of fossil fuels. The main source for the anthropogenic CO_2 emissions is coming from the consumption of non-renewable energy sources and non-sustainable land use practices.

The *Slow Carbon Cycle* is one of the "traps" for atmospheric CO_2. It dissolves CO_2 in rainwater and forms carbonic acid (H_2CO_3) and dissolves rock. Eventually the transport mechanisms take carbon to the ocean via rivers and where it is buried in the ocean floor in sediment. This process rebalances the Earth's temperature over geological time over hundreds of millions of years. Another process taking place over millions of years is tectonic and volcanic activity in which the tectonic plates on Earth's surface move and releases CO_2 through continental rifts and volcanic eruptions. *The ocean water-based carbon pump* is not quite as slow process as the *Slow Cycle*. In this process the ocean water absorbs CO_2 through diffusion and forms hydrogen and bicarbonate ions and makes oceans more acidic. The denser cold water that has the dissolved bicarbonate moves then the carbon deep into the ocean from which in time frame of thousands of years the carbon returns to atmosphere by warmer currents impacted also by water salinity and wind. Although increased CO_2 concentration improves this carbon absorption the raise in ocean temperature reduced the uptake more and amplifies further the greenhouse effect.

The *Fast Carbon Cycle* is based on biology and is also called *biological carbon pump*. In land the plants utilize the energy of the sun and grow by using photosynthesis. This reaction intakes CO_2 from the atmosphere, bicarbonate and oxygen. This process stores carbon in the plant's trunks, stems and roots. Eventually living organisms die, and carbon is released to soil where it may remain bound or may be decomposed by microbes and freed back to the atmosphere. The process of carbon release to atmosphere increases after temperatures are high enough to cause soil respiration and lack of water due to mainly evaporation leading to decomposition and thus carbon release. In terms of oceans the biological pump is based on phytoplankton and algae growth that captures CO_2 and uses photosynthesis. The produced hydrocarbons are eaten by ocean life and through decay ends up back to atmosphere or sinks to the ocean floor and forms sediment. The main role of this cycle is the movement of carbon from top to bottom. With increase in temperature this mixing is limited and therefore the lack of nutrients and oxygen can create dead zones and death of ocean life.

Aerosols from human activities in the atmosphere are also a source of greenhouse effect. Aerosol particles and typically suspended in a solid or aqueous form and in the range 0.001–10 μm in size (1 μm is 10^{-6} m). Due to their small size, they can move in the atmosphere easily and can remain suspended in there for days to weeks. They are normally anthropogenic; however, certain natural events also create them (e.g., volcanic eruptions). Main sources that we can influence and take responsibility for include burning fossil fuels and biomass. The greenhouse effect by aerosol particles depends much on their composition and location in the atmosphere. This connects to the fact that color and for example chemical composition may influence the absorption and reflection factors related to different wavelengths of solar radiation and heat. The most absorptive aerosols and therefore most impactful in terms of greenhouse effect are *black carbon aerosol particles* that are formed by combustion or separated from fuel during combustion and rise to the atmosphere. The opposite effect of reflecting solar radiation is with aerosol particles that come from volcanic eruptions and sulfur oxide and other sulfur compounds from burning coal. The force of volcanic eruption transports the sulfur-containing gases into the *stratosphere* (the second layer of the atmosphere). The sulfur dioxide that is coupled with other carbon emissions remains in *troposphere* (lower layer of the atmosphere) and adds to formation of aerosols that directly and indirectly has effect in warming and cooling in the atmosphere. Both are able to reduce the greenhouse effect. These aerosols have been proposed as a geoengineering solution to reduce global temperatures. Unfortunately, the side effects may easily cratc much worse problems.

Planetary orbit and solar variations are effects that change Earth's temperature in a cyclical manner. Our planets' orbit and its spin around its own axis are not constant. Also, the tilt of our planet's axis changes over time. The wobbling movement around the axis has a 26,000-year cycle and the planetary tilt changes from its low value to its maximum cyclically in every 41,000 years. Lastly the elliptical orbit around sun is slightly off and creates a solar energy variation during a few hundred-thousand-year cycles. These events may even overlap and can create cooling or heating of our planet that can impact carbon cycles and climate. Another, but shorter cycle is the 11-year solar cycle in which sunspots, flares and coronas changes energy output that can impact our average temperatures. These variations have however been reported to have only a fraction of degree impact on average.

2.4 Anthropogenic GHG Emissions

Let's look at the scope and basis for comparison for the anthropogenic CO_2 emissions. During the era of industrialization our emissions have been about 2300 billion tons of CO_2. About half of that has been dissolved through carbon cycles but the rest are left in our atmosphere. These emissions have thus increased the atmospheric CO_2 concentration by 65% (from 250 to 412.5 ppm). As a result of this significant increase, the average global temperature has also experienced a rise of ~ 1 °C since the pre-industrial era. Annual fossil fuel related emissions were projected in 2021 to be 36.4 billion tons per year. If we also include the land use change (deforestation, drainage and burning of organic soils) related emissions, which can give both positive and negative impact, the total CO_2 emissions was 39.4 billion tons per year (2021).

When discussing climate change related emissions, we have to be very careful with the details. The numbers we have been discussing are only the actual CO_2 related emissions and not the *CO_2 equivalent* (or CO_2-eq) which are used as a unit as well. The CO_2-eq numbers are used for the purpose of comparing the emissions of other GHGs on the basis of *Global Warming Potential* (GWP). The CO_2-eq numbers are utilized to facilitate the comparison of greenhouse gas emissions by considering their GWP. The approach allows for a standardized assessment of emissions from various greenhouse gases in relation to their contribution to global warming. This is done by changing from one form to another the quantities of other gases to the equivalent amount of CO_2 with the same GWP. Therefore, GWP is a metric used to measure the amount of energy that a specific GHG absorbs over a specific period of time in relation to the emissions of a ton of CO_2. The basic approach here is that CO_2 has GWP of 1 since it is used as the reference gas. However, methane has by estimation GWP of 28–36 (considering its impact over a 100-year timeframe), nitrous oxide has a GWP 265–298 also looking at impact over 100 years and the rest of the greenhouse gases, although they are in small quantities, have GWP in thousands of even tens of thousands. Finally, the GWP for water vapor is essentially not high enough to be quantified. It is not considered a significant factor in global warming as it is controlled by the water cycle and has a short (few days) atmospheric lifetime.

In 2019 the total greenhouse gas emissions, accounting all emissions reached about 60 billion CO_2-eq tons using a 100-year global warming potential time horizon (this is dramatically 40% more than in 1990s). Interestingly, the next year, due to global COVID-19 or SARS-CoV-2 pandemic, fossil fuel consumption and industry related emissions dropped by about 5% in 2020. This is the largest drop in absolute terms on anthropogenic CO_2 emissions on record. As a comparison, since the second world war similar relative emissions reduction has never occurred. These numbers have rebounded as the economic activity has revitalized and are currently back at pre-COVID level.

Putting things to perspective CO_2 emissions are about 70–80% of anthropogenic GHG emissions. Approximately 80% of CO_2 (CO_2) emissions are generated from the combustion of fossil fuels, which is used for various purposes such as heat and electricity generation, manufacturing processes, and transportation. Land-use change, and forestry are contribution about (2.2%) to CO_2 emissions and is mainly caused by deforestation. The other greenhouse gas emissions come from methane (17.2%), nitrous oxide (6.3%) and mostly from agriculture, waste treatment and gas flaring. The fluorinated gases with the high GWP come from industrial processes and account for about 2.3% of global emissions.

Reviewing different reports on sources of the global CO_2 emissions (e.g., "Our World in Data", "IPPC", "Circularity Gap") gives a good insight to where the highest impact and lowest hanging fruits for achieving results in terms of technology, research and development and financial investments (financial market or company investments) for carbon neutral solutions. The task of analyzing the different reports and viewpoints on the emission sources is not straight forward and it does not help that the numbers, definitions, and methodologies are numerous. Also, there is no single or simple solution to tackle climate change and therefore understanding its root causes is critical for impactful mitigation strategies. As the CO_2 emission categorization can be complex and highly depends on the detailed scope definitions it is good to look at several data sources and various different reports and literature.

A very powerful approach for exploring the climate change effect is to consider the connectivity to the current social needs. They unravel the connection to the modern civilization and the cost of keeping it a. This was reported by Circularity Gap Report recently (Circularity Gap Report 2022). The report is published by the Platform for Accelerating the Circular Economy (PACE), which serves as a collaborative plan of action and project accelerator to advance the adoption of circular economy principles. PACE operates through a partnership between public and private entities, working together to promote and support the transition towards a circular economy. It is also a collective with more than 70 leaders, and its foundation was initiated at the World Economic Forum.

The social need structure in the Circularity Report is based on seven categories: *housing, nutrition, mobility, consumables, services, healthcare, and communication.* Although the total emissions numbers used may not be entirely consistent to some of the other sources, they are in relative terms quite accurate. Their reported Greenhouse emissions were 59.1 billion tons of CO_2-eq GHGs in 2019. This number included so called Land Use Change and Forestry (LULUCF). The analysis in the Circularity Gap Report assesses the global economy through a viewpoint of circularity and by looking at GHG and materials, thus applying *"Mass-Carbon thinking"*. The premises for the analysis is to look at four distinct types of resources responsible for the GHG emissions by looking at the value chain or value created steps (i.e., Take, Process, Produce, Provide). The resource side shows clearly that fossil fuels represent the most global embodied emissions (65% or 38.4 billion CO_2-eq tons). In this category we have petroleum that fuels global transportation, coal and natural gas usage for electricity production and heat and powering industrial processes. The second largest GHG source embodied in a natural resource is biomass (27% or 16.0 billion CO_2-eq tons) from agricultural and forestry operations and processes. Minerals' extraction, processing and use count for only 2.7% or 1.6 billion CO_2-eq tons and ores only 2.0% or 1.2 billion CO_2-eq tons). Finally waste handling represents only 3.2% or 1.9 billion CO_2-eq tons or GHG emissions.

As these resources are going through a transformation process within their value chains there is a conversion from energy carriers to final products or services that society needs. The energy carriers and materials go through industrial manufacturing processes and are transformed to products made of plastics, metals, minerals, or wood. They all have been produced by adding process heat and electricity, together with various raw material resources (including products from petroleum refining), reused, or recycled materials or *waste valorization*. In the next step the final products and services are created, including necessities such as food and clothing, healthcare, building materials, and technical equipment. At this final stage, the balance has shifted, and all materials combined account for 51% of GHG emissions leaving energy use for heating, cooling, and lighting our workspaces and home and energy used for transportation to account for the remaining 49%.

Further analysis shows that from societal need perspective *mobility*, in general, is the highest contributor to GHG emissions with 28.9% (17.1 billion tons CO_2-eq). The majority of it is coming from the use of fossil fuels in the combustion engines with a much less portion from the actual automotive, trains and airplanes production. The second highest emissions come from *housing* with 22.8% of all GHG emissions (13.5 billion tons CO_2-eq) this high amount is connected to the associate use of enormous amount of material (38.8 billion tons) and related excavation, transport, and construction activities. In addition to this, housing has a high carbon footprint coming from energy used for heating, cooling, and lighting. *Nutrition* is responsible for ~ 16.9% of GHG emissions, which amounts to around 10.0 billion tons of CO_2-eq. This significant contribution is primarily associated with the high biomass weight of ~ 21.3 billion tons, which originates from food production activities. Additionally, the link between nutrition and GHG emissions is closely connected to Land Use, Land-Use Change, and Forestry (LULUCF) practices.

The *services* category includes public services such as carbon footprint for public facilities and offices, schools and material need for their operations. The *services* share of the total GHG emissions is 10.8% (6.4 billion tons CO_2-eq). The third lowest GHG emissions (3.5 billion tons CO_2-eq or 5.9% of total GHG emissions) comes from *communications* and is related to the infrastructure (networks, devices, and storage). Interestingly the second lowest contribution comes from the *consumables* and is only 9.5% of total GHG emissions (5.6 billion tons CO_2-eq). This category accounts for all manufacturing of consumer products including electronics, clothing, and personal care. Although, typically in high focus in mitigation of GHG emissions consumer products is not the highest impact category in the overall scheme of carbon neutrality. The last and least category for societal needs is *healthcare* and it represents 5.1% of all GHG emissions (3.0 billion tons CO_2-eq) and its emissions come from hospital facilities and buildings, healthcare technology and products and pharmaceuticals (Fig. 2.1).

Another point of view to the same analysis demonstrating the origin of the emissions come from, is to take a look at industry sectors and processes. 'Our

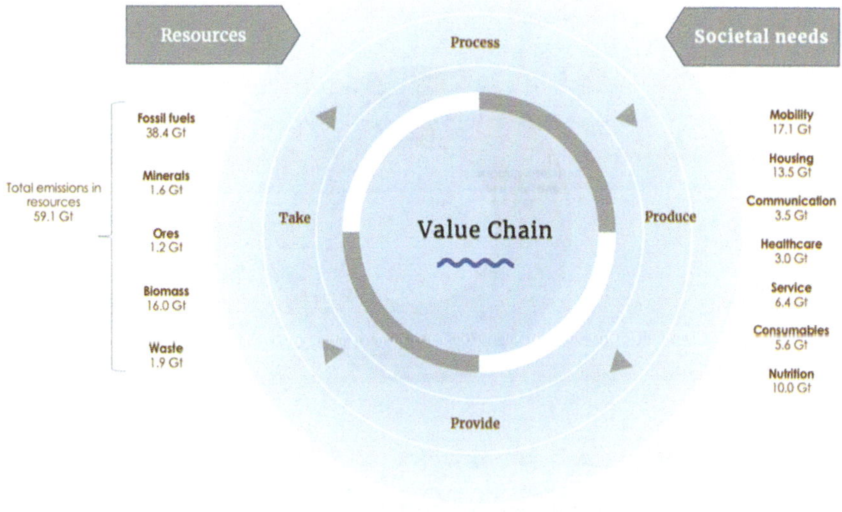

Fig. 2.1 Value Chain Presentation from Resources to Societal Needs

World in Data' is a project of the Global Change Data Lab, which is a registered charity in England and Wales. It functions as a research organization and creates data for the benefit of solving the world's largest problems. In their analysis on percentage of global Anthropogenic GHG emissions in categories such as *Energy use in Industry* (24.2%), *Agriculture Forestry and Land-use* (18.4%), *Energy use in buildings* (17.5%), *Transportation* (16.2%), *Industry chemicals and cement* (8.4%), *Waste* (3.2%), *Unallocated fuel combustion* (7.8%), *Fugitive emissions* from energy production (5.8%) and E*nergy in agriculture and fishing* (1.7%) (Climate Watch, 2022).

The total GHG emissions in 2020 in Climate Watch analysis was 49.4 billion tons of CO_2-eq and the categorizations are quite different compared to the Circularity Gap Report. For example, in Energy use in buildings, by their definition, does not include the related excavation, transport, and construction activities. Also, in the Circularity Gap Report the building related emissions are distributed by use in societal needs categories. For instance, building emissions in hospital*s* are in *healthcare* category and carbon footprint for public facilities and offices are in the *services* category (Fig. 2.2).

Another category in the Circularity report that needs some attention is biomass. Although the 16 billion tons CO_2-eq is quite close to the *Agriculture Forestry and Land-use* category of the 'Our World in Data' report, there is a difference. As the *biomass* collected goes through the value chain the emissions are allocated to other categories such as *consumables* for clothing, mobility for *transportation* and biofuels and some biomass composites and fibrous materials used in automotive industry.

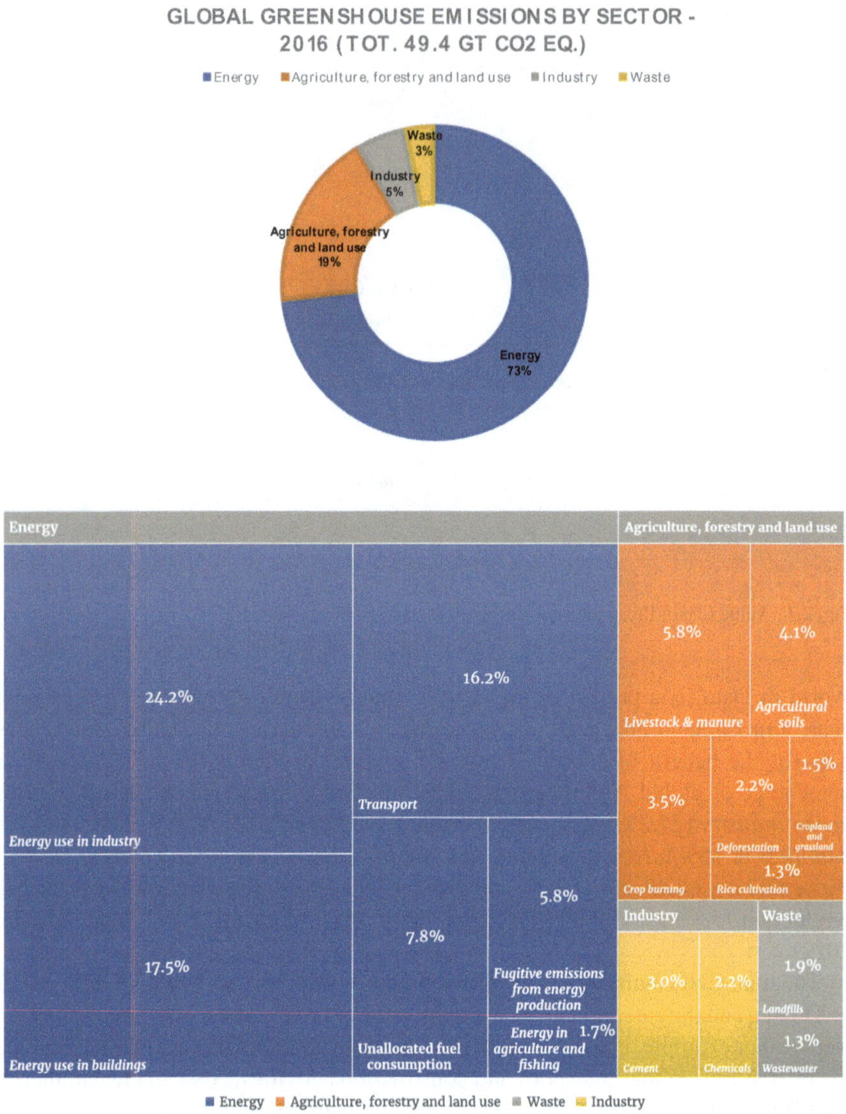

Fig. 2.2 Greenhouse gas emissions by industrial sector

Lastly the nutrition is the largest contributor as a societal service coming from the *Agriculture Forestry and Land-use.* The category Transport (in the 'Our World in Data') only includes the emissions from electricity produced or burning fossil fuels to power transport activities. The definition does not include emissions from manufacturing any vehicles or equipment. Those numbers would be included in the numbers of the *Energy use in Industry* category. The list goes on and for the reader the complexity can be unveiled by following the value chain and end use or purpose.

In order to see more industry sector related perspective, we should also focus only on the energy production using IPCC (2014) and 'Our World in Data' data. The total energy use accounts for 73% of the GHG emissions. More specifically, the *electricity and heat production* accounted for one-fourth of the global GHG emissions (IPCC, 2014). It is coming from usage of coal, natural gas, and oil and also includes the industrial electricity use. Additionally, there are also fugitive emissions from energy production and fuel extraction and refining processes, and transportation related emissions (10%).

The *manufacturing industry's* use of energy was creating 21% of total GHG emissions in 2010 (IPCC, 2014) and involved the fossil fuels used on site in industrial manufacturing and other facilities. Furthermore, this sector encompasses emissions resulting from chemical, metallurgical, and mineral processes that are not directly associated with energy consumption. Additionally, it encompasses emissions generated from diverse activities within the realm of waste management. A more detailed examination ('Our World in Data' 2021) shows somewhat similar numbers (total 24.2% of GHG emissions) but more importantly *Energy use in Industry* and the associated GHG emissions is related to various industries and their processes including iron and steel industry (7.2%), non-ferrous metals (0.7%), machinery (0.5%), food and tobacco (1.0%), pulp and paper (0.6%), chemical & petrochemical (3.6%) and other industry 10.6%.

Accordingly, a major energy use and GHG emissions related to industry is also connected to *Transportation* (16.2%) and predominantly includes the use of fossil fuels in personal vehicles or trucks (11.9%), aviation (1.9%), railroad (0.4%), maritime transportation (1.7%) and fuel or gas transport using pipelines (0.3%). The energy use in buildings (total of 17.5%) comes from residential (10.9%) and commercial (6.6%) as also adds to the industrial GHG emissions.

Somewhat different viewpoint in looking at the big picture of the GHG emissions reporting from a global perspective comes from seeing how the different gases contribute to the whole global emissions. According to the IPCC's (2014) report, ~ 65% of total CO_2 emissions are attributed to fossil fuel combustion and industrial processes. An additional 11% of emissions are associated with forestry and land use operations. This would include deforestation and clearing of land for agriculture, and soils degradation. This estimate considers the removal of CO_2 from the atmosphere through activities such as reforestation, soil improvement, and other related practices. These activities help to offset a portion of the CO_2 emissions and contribute to overall carbon balance. Methane is a large contributor to the GHG emissions (16%) and can be sources to agricultural activities, biomass burning and waste management. More GHG agricultural emissions come from nitrous oxide from fertilizers, additionally it also comes from fossil fuel combustion. In total this makes up for 6% of GHG emissions according to IPCC. The final category comprises fluorinated gases, which account for ~ 2% of greenhouse gas emissions. These gases are generated through various industrial operations and processes, such as refrigeration and the use phase of certain products. Their release into the atmosphere contributes to the overall emissions of greenhouse gases.

There is growing recognition of the *GHG emissions linked to military operations, wars, and conflicts*. One recent area of concern is the impact of Russia's war in Ukraine, which is undermining global efforts to address the climate crisis. Similarly, the armed conflict between rival factions of the military government in Sudan is offsetting progress made in other parts of the country.

According to an assessment by de Klerk et al. (Climate Damage Caused by Russia's War in Ukraine, 2022), the GHG emissions resulting from seven months of war in Ukraine are estimated to be around 100 million metric tons CO_2-eq. To provide perspective, this is roughly equivalent to the GHG emissions produced by the Netherlands over the same period. The analysis draws on various sources, including scientific articles, intelligence reports, and satellite imagery. As the war is ongoing, it continues to have adverse effects that hinder global efforts to mitigate GHG emissions.

War generates GHG emissions from multiple sources. Fossil fuels play a significant role due to their use in armored vehicles, tanks, aircraft, and other military machinery. Artillery munitions and fires resulting from bombing and mine-laying operations are additional sources. The damage to civilian infrastructure, which also factors into the post-war reconstruction of that infrastructure, accounts for approximately half of the total GHG emissions during wartime. Fires caused by the conflict are another significant contributor. Emissions related to the transportation of refugees and internally displaced persons (IDPs) are relatively minor in comparison.

Overall, the GHG emissions associated with military activities and conflicts underscore the need to address the environmental impacts of warfare and strive for sustainable and peaceful resolutions to prevent further harm to the climate and the environment.

2.5 Thoughts on Value Chain Improvements

Now taking these analysis makes us able to conclude what are some of the high impact areas in GHG mitigation from a global perspective. Using the societal needs categories, *mobility* has clearly the highest impact to GHG emissions with 28.9% (17.1 billion tons CO_2-eq) of all global GHG emissions. Increasing efficiency of vehicles (aircraft, car, trucks, and trains), is a critical factor in GHG mitigation, unfortunately the turnover of vehicle fleet and especially trains is typically more than 10–20 years. Logistical efficiency improvements can, however, be facilitated by the latest information and communications technology (ICT) and artificial intelligence (AI). The electrification of transportation is clearly a major opportunity together with increasing the use of lower-carbon fuels. Public transportation is societal and relates to city planning and provides a low emissions alternative to commute and for general transportation. It also facilitates the development of more efficient and compact communities with shorter distances.

The *housing value chain* is estimated to be the second largest contributing factor to GHG emissions with 22.8% of all GHG emissions (13.5 billion tons CO_2-eq). It includes various sub processes that add to the energy use and thus emissions. If we look at the very first steps in the value chain, that is the mining and extraction of minerals, the mining industry is able to decarbonize or reduce GHG emissions through various approaches. Some of the approaches require major new capital investments, however, many solutions such as increased use of renewable energy, electrification, and operational efficiency, are economically feasible even short term in most of the current mining operations. For transport of construction materials there are three routes to reducing GHGs increasing the efficiency of vehicle technology, changing how the materials and goods are transported, and using lower-carbon fuels. Efforts to reduce GHG emissions in the construction sector can focus on improving the efficiency of material design. This can be achieved by minimizing material waste, increasing the use of recycled content, and finding alternatives to high-emission materials like steel and cement in construction projects. By adopting these measures, the construction industry can contribute to reducing its carbon footprint and promoting more sustainable practices. This can be improved by expanding the use of low carbon cement or sustainable timber or for example Cross-Laminated Timber (CLT). In addition, increased building utilization efficiency, including use of heat pumps for both cooling and heating and low energy light-emitting diode (LED) lighting are some of the approaches to create low energy housing by an energy-efficient technology design. Zero energy building is the latest concept combining energy efficiency and use and generation of renewable energy. It aims at consuming only the energy that is produced onsite using renewable energy resources.

The nutrition or food industry value chain is a significant contributor to greenhouse gas (GHG) emissions, accounting for ~ 16.9% of total emissions, equivalent to 10 billion tons of CO_2-eq. This highlights the substantial role that the food industry plays in contributing to climate change-related emissions. Addressing emissions throughout the food value chain, from production to consumption, is crucial for mitigating the environmental impact of the sector and achieving climate goals. Interestingly its impact has been highly underestimated until recently. The total GHG emissions have been rising about 8% between 1990 and 2018. However, the per capital emissions have been trending down. The industrial nations still have a lot to work on as their per capital emissions are twice the emissions the developing countries have. Increasing crop and livestock productivity is an apparent approach to mitigate emissions, however, we also need to consider that the farmers should not convert natural forest or grassland to cropland. Funding and governmental policies are needed to conserve carbon-rich natural landscapes.

Interestingly the production of livestock accounts for almost half of the GHG emission in food industry. The main GHG sources in livestock management are *enteric fermentation* (or cow burps), manure and waste management. In addition, emissions from animal feed production (e.g., corn, soy) can also play a significant role. There are new feed mixtures that are designed to improve the animal ingestion and thus reducing the methane produced. The other major source for emissions is fertilizers (synthetic and animal manure). Some of the existing solutions for nitrogen oxide emissions from synthetic fertilizer include nitrification inhibitors that slowly release the nitrogen for the crops and thus reduce the unnecessary over fertilization over longer period of time. Also, microbes that enable the crops to utilize their own nitrogen more efficiently, have been developed. Of course, new policies are needed including establishing new mandatory standards for fertilizer companies and users to enforce more efficient use of nitrogen in their fertilizers. Other approaches to GHG emission reduction includes improving the efficiencies in fuel consumption in farming equipment. Also, more energy efficient buildings in farms and the use of self-produced and access to other alternative renewable energy systems are needed. Lastly, the reduction in food loss and waste in the supply chain and by consumers and restaurants can have a large impact to the overall emission from the *food industry value chain*. This can be achieved for example by technologies improving food shelf life and logistics.

The *services category* comprises of public services including offices, schools and material that is needed for running them. It has a 10.8% share of the total GHG emissions (6.4 billion tons CO_2-eq). An important approach to mitigate climate change in this category is the use and adaptation of the Leadership in *Energy and Environmental Design* (*LEED*) green building rating system. The LEED framework drives for high energy efficiency and cost-savings in green buildings and assists buildings in reducing their GHG emissions. This approach also addresses sustainability from materials selection, air quality and human health and wellbeing perspective. LEED certified buildings have been reported to contribute 50% fewer GHGs due to lower water consumption and 48% less due to improvements in solid waste management (Mozingo and Arens 2014). Other building related improvements could include the use of use of heat pumps, LED lighting and improved insulation in the building maintenance and repair schedules. Public services materials use can be affected by the public procurement policies and therefore by making sure that sustainable materials are preferred.

Although the *value chain for consumables* do account for as high as 9.5% of total GHG emissions (5.6 billion tons CO_2-eq), the public perception is that it would be even more. Consumables are a commodity that is used by individuals and businesses and must be replaced regularly because they wear out or are used up. In this category food is not included. Some examples of consumables include plastics products, clothes and textiles, paper products, electronics, books, and furniture. Another way to reduce the associated GHG emissions in this category is to use *Sustainable Design* practices for product development. More specifically replacing the single-use plastics with paper-based products, designing longer lasting electronics devices to minimize e-waste, and using recycled plastics, natural fibers, wood, and related by-products are some of the approaches to reduce emissions. Other things to work on include improving and increasingly introduce reuse and recycling business models so virgin material usage is less. Improving product quality and lifespan is important. When it comes down to manufacturing improved efficiency and continuous reduction of chemical use or replacement of petroleum products by using biobased alternatives is critical.

The *Communications* value chain is responsible for 5.9% of total GHG emissions (3.5 billion tons CO_2-eq) as such driving down emissions in the in this category comes through the development more efficient information and communication equipment and networks. For example, if consumers were buying more laptops instead of desktops with monitors this would reduce the GHG emissions. E-books have already and continue to reduce the use of paper and is reducing the need for transportation of heavy books. Smart phones consume quite a small amount of energy to operate, however, the production phase is responsible for 85% of their total emissions. A smartphone's chips and motherboards include precious metal with high GHG associated to their mining phase. In addition, the life span of smartphones is quite short, and we need to keep in mind that there are already 6.5 billion smartphones in use globally. A large portion of the GHG emissions contribution come also from the telecommunications infrastructure, which mostly comes from the data centers that enable modern cloud-based internet and telecommunications services. Some mitigation approaches for high carbon footprint of the data center include the use of outside air and water for cooling instead of air conditioning and powered refrigeration. Also, renewable energy sources, including solar panels and wind power, can be used in addition to advanced temperature controls. In addition, *Infrastructure Management Tools* (*IMT*) can be used to improve energy efficiency by analyzing data and optimizing data center design, energy management and capacity planning.

The last category we will discuss from very general perspective is *healthcare*. It represents 5.1% of all GHG emissions (3.0 billion tons CO_2-eq) and maybe considered as small compared to the rest. Still, healthcare emissions have similar development needs as *housing*, *communications*, *services*, and *consumables*. That is using the technological assets long will reduce the GHG emissions over a lifetime. Travel needs can be reduced by adding virtual doctors' visits, and hospital consumables including surgical masks, gowns, and disposables can be designed and manufactured more sustainably. The hospital facilities can be LEED certified and can use more energy efficient cooling and heating systems and renewable energy for their operations.

2.6 Hard to Solve Carbon Neutral Challenges

Many of the above-mentioned general level or conceptual GHG mitigation topics and approaches seem to only be missing economic incentive and proper policies. Later on, we will discuss in more detail some of the technology platforms that do create solutions and also the research and development that is needed for an effective and high impact techno-economic solution. As we try to keep track of focusing on the most critical and high impact approaches for carbon neutral solutions, let's now look at some of the hardest emissions to eliminate. About 25% of the global GHG emissions are estimated to be cumbersome to reduce or eliminate. These include *reliable and load-following electricity* with (14% of total GHG emissions), *steel production* (5% of total GHG emissions), *cement production* (4% of total GHG emissions), *shipping—rail, ships, other* (3% of total GHG emissions), *aviation* (2% of total GHG emissions) and *long-distance road transport* (1% of total GHG emissions) (Davis et al. 2018).

The *high reliable and load following electricity* is critical to all modern societies. As we increase renewable energy supply this is increasingly important due to the fluctuating availability of solar or wind power. Natural gas-fired generators or turbines and other fossil-based fuel sources are an example of an affordable approach to generate he necessary flexibility. Recent research by Davis et al. (2018) has estimated this flexibility to contribute about 12% of nonrenewable energy and industry emissions. Regardless of the sustainable energy mix, even with nuclear or coal-burning power plants with carbon capture and storage (CCS), the same energy system dilemma exists and therefore generators such as gas-fired turbines with high availability and variable cost are always needed. A potential solution is to use a mix of natural gas, synthetic hydrocarbons (syngas) or hydrogen as fuel in combination to CCS. This would make the system sometimes even carbon negative. Unfortunately, the capital costs of these solutions are still not feasible. Nuclear power can also be used for load following but it demands additional investments. New concepts are currently being developed and can potentially create feasible fission energy solutions to this problem. Energy storage technologies are also critical in achieving reliability. For example, stored hydrogen (green hydrogen from electrolytic production) can be used by fuel cells or by gas turbines. One other venue is to use hydrogen by combining it with CO_2 via methanation (renewable methane) or it can be mixed by using lower concentration ($< 10\%$) of natural gas or biogas for energy and electrical power generation. Battery technologies can also be used for short-duration reliability improvements, the provided discharge electricity cost is easily 5–10 times the base cost of electricity. Pumping water to reservoirs is another mature and economically feasible option. The main negative issues include finding available water and suitable reservoirs, and the social and environmental risks. Underground and undersea design options may however be solutions for mitigating these obstacles and risks. Energy can also be stored using underground formations or pressure vessels and transformed back to electricity by using turbines. Thermal storages include solid or molten materials (bricks, gravel, molten salt) and water tanks. Also, latent heat-based solutions can be used by using phase-change materials. This technology is, however, more suitable for a timeframe of day or so shifting of heating and cooling loads. Demand management and intelligent charging of EVs and utilization scheduling of appliances can also create peak load reductions that are helping with reliability, however, the full scope solution to *high reliable and load following electricity* is a portfolio of these various technological approaches. They have to be economically feasible and reliable. The reduction of capital costs is therefore always critical and drives innovation and technology diffusion and deployment.

Infrastructure and inhabitant construction is part of industrialization and also has a major contribution to GHG emissions. The two most energy consuming, and CO_2 intensive construction materials are steel and cement. The *production of steel,* which relies on the use of coking coal as a fuel and reducing agent in blast furnaces, contributes to ~ 6% of global CO_2 emissions. This highlights the significant carbon footprint associated with steel production processes. Finding alternative methods that reduce reliance on fossil fuels and promote more sustainable practices is crucial for addressing emissions in this industry and achieving climate targets. In steel production process there is about 1.85 tons of CO_2 produced per ton of steel produced. The production temperature is high (~ 1600 °C) and therefore also demands a lot of heating energy that also produces GHG emissions. One of the carbon neutral solutions to this high GHG emitting industrial process is to use electric arc furnace (EAF) technology. This technology in combination with renewable electricity production, improved efficiencies and direct reduction process can make improvements to the GHG emissions. Some other approaches include using biochar instead of coke and hydrogen as the energy source or as a reductant either as auxiliary reducing agent or as direct reduction of iron (Yilmaz et al. 2017; Patisson and Mirgaux 2020). CCS technology in combination with furnaces that use higher CO or CO_2 concentrations can also be used.

The *manufacture of cement* emits a considerable amount of CO_2. About half of the emissions come from the fossil fuel-based energy used in the process. The other half comes from the calcination of calcium carbonate ($CaCO_3$) when it is heated to 600–900 °C in a rotary kiln and goes through complex chemical reactions. This conversion to lime (CaO) or clinker, the primary component of cement, releases CO_2 as a by-product. Part of the kiln is in a higher temperature and makes CaO react with silica, iron, and aluminium containing materials producing the precursor to cement. After cooling and grinding it is mixed with gypsum creating the well-known Portland cement. In order to reduce the carbon emissions, the whole process needs to be redesigned or a carbon-capture (CC) technology needs to be installed. The CO_2 concentration from the rotary kiln is high (~30%) and that makes the use of CC technology more efficient. Interestingly, cement in a long-time scale (50 years or more) does reabsorb a large fraction of CO_2 via natural carbonation process. This process can take up even 60% of the CO_2 that was released by calcination (Ravikumar et al. 2021). Therefore, in a sense if CC technology is utilized the there is a net-negative emissions involved in the production of cement.

In total, transportation accounts for 16% of global CO_2 emissions. *Air traffic, railroad, and shipping* represent about 4% of this total. The *short-distance light-duty road transport* (essentially personal vehicles) and *short distance medium and heavy road transportation vehicles* accounted for 12%. The reduction of such emissions is already ongoing with the explosive expansion of EVs during the recent years. Global sales of electric vehicles have reached 10% of all new cars sold in 2022. Compared to 2020 numbers in 2021 the sales nearly doubled. This electrification will continue exponentially as most automakers are committing to zero emissions. The challenge for drastically reducing CO_2 emissions in terms of aviation, maritime ships and long-distance transportation is, however, much more difficult as the necessary space for cargo and general weight aspects demand higher density power sources than batteries are able to provide. A matter-of-fact heavy-duty truck using state of the art battery technology would be able to transport 40–50% less cargo than their internal combustion engine-based counterparts. This same technological bottleneck applies to the aircrafts and marine transportation. As the liquid fuels seem to be the only option for the long-range transportation the options to reduce the current emissions rely on novel sustainable liquid fuels including green hydrogen (made using renewable electricity), ammonia and alcohols and ethers. All these options can be part of a net-zero energy system. The main challenge with hydrogen is that in order to achieve adequate energy density it needs to be stored in a liquid from with low temperature (−252.8 °C at 1 atm) and high pressure when in gaseous from (350–700 bar). These conditions require high capital investments, safety procedures and due to the high weight of such systems the total energy density is inferior to internal combustion engines. Similarly, ammonia is an attractive alternative as a liquid fuel, but it also has relatively low gravimetric energy density (Wh/kg or MJ/kg). Some recently developed technological concepts combine both liquid hydrogen and ammonia with stored electric power and there already exist a proof of concept for this solution that is from a performance wise competitive with traditional diesel trucks. Biofuels has been successful renewable energy source for transportation industry, they represent about 4% of all transport sector energy consumed. Biofuels include ethanol from corn and other grain sugar cane, biodiesel from plant seeds and waste oils. Biofuels have their pros and cons, with biggest challenges in scalability and cost. Also, the full scope life cycle emissions analysis and conversion efficiency of the biofuels can be somewhat disputed.

The latest advanced biofuels developments have included other woody biomass from various crops and residues and waste. This has reduced the competition on agricultural land use and when combined with CCS technology it can be used as a negative CO_2 emissions pathway further increasing the economic feasibility and reaching the same cost level as the fossil-based fuels. Other approaches include industrial Fischer–Tropsch process (originates to 1920s Germany) used for carbon monoxide to react with hydrogen under certain conditions and creating wide range of long-chain hydrocarbon products such as biofuels. This would create also truly carbon neutral fuels with CO_2 removed from a point source or the atmosphere by CCS technology. There is a lot of momentum in hydrogen-based solutions, but the feasibility of using green hydrogen in any of these approaches is dependent on the capital cost the electrolysis and the cost of renewable electricity. New technological approaches have continuously been developed and are currently being tested. In general, the renewable liquid fuels still remain one of the most potential solutions in combination with electrification for achieving scalable and economic net-zero emissions transportation energy system.

References

Circularity Gap Report (2022) https://www.circularity-gap.world/2022#Download-the-report. Accessed 15 March 2022

Climate Damage Caused by Russia's War in Ukraine (2022). https://climatefocus.com/wp-content/uploads/2022/11/ClimateDamageinUkraine.pdf

Climate Watch (2022). Washington, D.C.: World Resources Institute.
Davis SJ et al (2018) Net-zero emissions energy systems. Science 360:1419
IPCC (2014) Climate change 2014: synthesis report. In: Pachauri RK, Meyer LA (eds) Contribution of working groups I, II and III to the fifth assessment report of the intergovernmental panel on climate change Core writing team. IPCC, p 151
Mozingo L, Arens E (2014) Quantifying the comprehensive greenhouse gas co-benefits of green buildings. https://escholarship.org/uc/item/935461rm
Our World in Data (2021). https://ourworldindata.org/co2-and-other-greenhouse-gas-emissions
Patisson F, Mirgaux O (2020) Hydrogen ironmaking: how it works. Metals 10:922. https://doi.org/10.3390/met10070922
Ravikumar D et al (2021) CO_2 utilization in concrete curing or mixing might not produce a net climate benefit. Nat Commun 12:855. https://doi.org/10.1038/s41467-021-21148-w
Yilmaz C, Wendelstorf J, Turek T (2017) Modeling and simulation of hydrogen injection into a blast furnace to reduce CO_2 emissions. J Clean Prod 154:488–501. https://doi.org/10.1016/j.jclepro.2017.03.162

Chapter 3
Policy and Governance for Climate Change–Global and Local Approach

Abstract Policy work is a complex and diverse field, particularly on a global scale. In this chapter, we examine fundamental policy concepts and look into examples at both local and global levels. We explore environmental policy focused on climate change and carbon reduction targets, including the Greenhouse Gas Protocol, Science Based Targets, and the EU Taxonomy. We also briefly discuss the key climate change policies in the USA, EU, and China. Additionally, we discuss corporate policies and targets, delving into their origins. In this chapter, we also present insights from company executives and experts, providing different perspectives on the challenges related to corporate policy and target setting.

Keywords Policy and Governance · Global GHG policy · Greenhouse gas protocol · Science based targets · EU taxonomy · Corporate Policies and Targets

3.1 Introduction to Policy

Since policy work is complex and multifaceted, especially in a global setting, it is good to start with the basics. Policies, in general, refer to courses of action that are proposed by an organization or authority. In order to simplify this topic, we are using four main types of policies: public, organizational, functional, and specific.

Public policy includes actions, regulatory measures such as laws, and funding priorities on topics stated by a governmental body. Public policy is normally embedded in constitutional directives, legislative acts, and judicial decisions, and they are designed and regulated mostly in a country level (Knill and Tosun 2020). The public policies give direction through legality activities (e.g., legal guidance and ensuring compliance with applicable laws) including for instance economic issues (e.g., taxes and government budget), health (e.g., public healthcare and insurance), criminal justice (e.g., crime prevention and enforcement of criminal sanctions), education (e.g., early childhood education and university system), and environment (e.g., emissions control and water quality). In addition, there can be many other

M. Hakovirta, *Carbon Neutrality*, Springer Climate,
https://doi.org/10.1007/978-3-031-45202-4_3

specific policies such as energy policy (e.g., energy production and distribution), agricultural policy (e.g., subsidies and export control), and foreign policy (e.g., security and international order).

Organizational policy is something that is created for and by a specific organization (Wergin 1976). The organization's stakeholders are typically part of the process with a purpose of setting goals and targets for the activities of the whole organization. The organization leadership is generally considered as the policy administrators and therefore they in principle define an organization policy. This is done by working according to the overall governance model that the organization is using. Some examples include code of conduct, health and safety policy, and antidiscrimination and harassment policy. Policies such as these are by and large an indication of the organizational culture and follow its value system and the purpose the organization serves.

Functional policies are more specific to the organizational models. It is generally agreed that there are four basic organizational model types that are typically used both in private and public sectors. Some additional variety of different approaches also exist. Divisional organization is composed of separate units, offices, or entities in which the different functions operate in and that are needed to produce the products and/or services offered (Wies 1994). Functional organization on the other hand is composed of all the entities that are required to produce the products and/or services offered, and process organization is different as it defines the way (i.e., processes) the organization develops and delivers the services or products offered. The last main organizational structure is matrix organization that in principle organizes people and resources by both function and product (Weekes 1981; Kashani 2009). The other different approaches to structure an organization include market-based divisional and a variety of matrix structure-based solutions. When it comes down to functional policies, they can include financial, R&D, engineering and production, marketing, and other aspects related to specific functional activities. In large organizations, typically with complex organizational structures, separate units or departments may have such a unique role that the same organization has different functional policies for the same activities. They of course need to follow the overall high-level policies and cannot be conflicting with each other.

Specific policies are an exception to the rule and are formulated and implemented for a particular purpose and for a limited duration. They exist in local and country levels, in public and private sectors. Examples of governmental-specific policies include such as activities that are considered at certain time dangerous or a threat to national security. A recent example is COVID-19-related policies that have changed every few months and are addressing citizen behavior, vaccination mandates, or international border control. In the private sector, for example, travel policies may change based on the financial situation of the company, or in public sector, this could include procurement policies that maybe timing dependent (due to trade embargo or other restrictions).

Global policies are a crucial part of global governance. Arguably, some of the most critical global issues affecting us all are social, political, economic, and environmental issues. All these matters are mostly interrelated and therefore influence each other. These global megatrends-related issues clearly connect with several topics including global population increase and urbanization. For example, increased wealth and standard of living demand more industrial activity such as manufacturing, which consumes more energy and fossil fuels, thus speed up climate change. These and many other global trends and issues are multifarious by nature. On many occasions, they do not obey national boundaries and local policies do not protect us from their worldwide impact. Recurrently, we experience global and regional dynamics that have a selectively different and unfair impacts and outcomes in different parts of the society and the world. Unfortunately, those nations and parts of society that are deprived and financially weak suffer the most.

The global governance manages complex issues and engages stakeholders on collective problem solving. It comprises different institutions, policy mechanisms and processes among different states, markets, and other organizations. Examples of the global governance institutions and bodies for regional coordination include the United Nations (UN), Organization for Economic Cooperation and Development (OECD), World Trade Organization (WTO), North Atlantic Treaty Organization (NATO), International Criminal Court (ICC), European Union (EU), and Association of Southeast Asian Nations (ASEAN). These and many other entities engage governments, international organizations, and the global business community for partnerships and for policy dialogue and knowledge transfer. They also coordinate transnational issues and facilitate cooperation and try to resolve any possible disputes between global actors.

3.2 Environmental Policy—Basics

The general approach to *environmental policy* is an extension and a commitment by the governmental body, public or private organization, or corporations to follow responsibly policies, laws, and regulations regarding environmental issues. And more specifically in a preventive and reductive way to minimize or eliminate the effects of human activities on the environment. The environmental policy issues mostly include air and water pollution, waste management, and the management of natural resources including wildlife, and threatened species. The principles of these policies and related laws are quite universal and rely on precaution, prevention, and remedying pollution at source, and on the enforcement of "polluter pays" assumption.

The reason for the importance of the existence and reinforcement of environmental policies is related to the fact that anthropogenic environmental impacts are in principle economic externalities. The cause and effect are not in most cases clear and consequences of creating pollution are seen elsewhere and far in the future. This is also associated to the theory of the "tragedy of commons." It states that individual actors only look at their self-interest and work against common good of all users. This creates depletion of resources, emissions, and pollution due to the lack of coordinated actions. The concept originally was introduced in an essay written by the British economist William Lloyd in 1833 and became more popularly known in 1968 when American ecologist Garret Hardin introduced it. This thought of using common resources with no consideration of limits is clearly preindustrial. In a global economy that is highly industrial, resources are scarce, and emissions and pollutions are imminent. The seemingly beneficial short-term view to environmental impact and sustainability is a cost that the society pays in the long run. The lack of incentives to operate sustainably must therefore be compensated by the local and global environmental policies.

The governance of environmental policy and management has changed during the past few decades (Environmental Policy 2020). It has transformed from a top-down government and public sector approach into a broader, holistic, and inclusive activity with high public interest in which communities, NGOs, and companies are involved in a more responsible way. These changes are due to the overall shift in governance models that many countries are realizing and moving towards more engagement with the private sector and civil society. What drives this change is the fact that the complexities in society have increased due to factors such as industrialization, globalization, and technological advancement. For these reasons, policymakers use a range of different environmental policy mechanisms. We can categorize these mechanisms to *regulation, market-based instruments (MBIs), voluntary methods, education, and communication*. The traditional approach is the *regulation*, which yet continues to be the backbone for environmental policy in most nations. Regulations are considered as exercise of authority and direction (command-and-control) instrument. Generally, this means any federal, state, or local law, code, municipal ordinance, or rule that is related to toxic or dangerous and hazardous pollutants, substances, materials, and waste. Examples of this for instance in the USA would include the Clean Air Act and the Clean Water Act. Unfortunately, regulatory mechanisms, although mostly used, have not been fully successful in what they were designed for, partially due to the restrictiveness, contextual representation, and lack of interpretation. The *MBIs* are policies that are market-based, and their use is designed to create incentives to reduce or eliminate environmental externalities, such as pollution and emissions. Examples of these market-based policy instruments include taxes, charges and fees, certain subsidies, emissions trading, permit systems, and deposit-refund systems. The MBIs apply both to consumers and producers, and there is in general a broad acceptance in using MBIs, but they have also been under criticism due to equity issues and some inefficiencies in resources allocation due to, for example, market failures.

Voluntary methods in environmental policy are quite different and can be highly important as there can be situations in which they are the only option on hand. This scenario can arise when there is no effective government body that can enforce or create environmental policies, or when there are shortcomings in the oversight and enforcement processes. Voluntary methods can include schemes where public and private participants are voluntarily participating in reducing pollution and emissions and committing to environmental improvement. These can be cross-industry efforts that are aiming at global impact such as climate change mitigation. Or they can be more specific efforts and address smaller scale and defined industry sectors, initiated by, for example, industry using unilateral commitments. Involvement may include local government in state, county, and city levels.

Most voluntary methods have detailed environmental goals but also programs that are designed more for raising environmental awareness are used. Communication in this respect is in a key role including marketing of some of the environmentally responsible products the participating companies offer. Some of the most challenging aspects involved include the lack of resources in volunteer-based programs and the overall lack of strategic direction. The educational policies are about citizens learning and exploring global and local environmental issues and getting them involved with problem solving and activities to improve the environment. Their role in environmental policy work is critical as participants are able to improve their understanding of the environmental, and also, they build knowledge and competence for responsible decision making both within the communities and in the private sector during their careers.

All these environmental policy mechanisms are in some way connected as for example there is self-regulation in voluntary approach, and BMIs can also include regulation and communication and education as part of successful implementation. It should be emphasized that the policy mechanisms need to be aligned with certain environmental, societal, political, and economic situations. Also, the overall understanding of the situation and the attitudes, competencies, and capabilities of the stakeholders (industry, citizens, public sector) is critical. With the purpose of achieving a successful design and implementation of environmental policy, the cooperation of all stakeholders needs to exist. In addition, open dialogue, mutual understanding, and general agreement of the critical issues are required. And it is essential to understand how policy instruments and working with the government can impact positively the environmental situation at hand. In all these mechanisms, the main forces that influence the success of policy implementation are connected to the realization of similar policies in the past and elsewhere. Also, the pressure of international community and organizations and the dynamics of other policies that may compete with the resources needed for the new policy need to be well understood and considered.

3.3 *Global GHG Policy* Introduction

As a broad introduction to GHG-related policies, the global authority for climate data is IPCC. They are highly respected and accepted by the various stakeholders and policymakers in climate change. IPCC produces comprehensive Assessment Reports (ARs) about climate change, its background and impact globally. Many of the reports have been on findings and estimates on the impact of 1.5 °C increase in global temperatures. In addition, they produce reports on specific issues and on various methodologies and guidelines including how to formulate GHG inventories (list of emission sources and the associated emissions). Although IPCC publishes reports every year, the more comprehensive and wide-ranging scientific ARs are published in every six to seven years. The latest AR6 was published in 2021, and the previous AR5 came out in 2014. If we look at the current policy work, the 1.5 °C global temperature increases, and its impact can be seen common as a foundational basis for all policy work. Other foundational work for GHG policies includes the Paris Agreement of December 12, 2015. The agreement is legally binding and was originally ratified by 193 parties, with world's major economies as participants: USA, China, and European Union. Some may argue that the agreement is more, or less voluntary, or a hybrid of regulatory and voluntary policy, as penalties such as fees or embargos, for those countries that are not able to fulfill their commitments really do not exist. There is also no international court or strictly enforcing governing organization that makes sure the compliance with Paris Agreement exists. Another important organization connected to global GHG policy is the United Nations Framework Convention on Climate Change (UNFCCC). They are at the heart of the Paris Agreement since through their process each participating country is required to establish and submit Nationally Determined Contributions (NDC) and update it every five years. These disclosures are then recorded, and each participant is held accountable to their committed action plans. The world's largest economies (US–GDP $23 trillion in 2021, China GDP $18 trillion in 2021, and EU–GDP $17 trillion in 2021) are also committed to these plans and have their NDCs submitted for the Paris Agreement.

In the United States, the largest global economy, President Joe Biden's executive order, issued on December 8, 2021, establishes climate policy objectives. Among these goals is the federal government's commitment to achieving net-zero CO_2 emissions by the year 2050. This is aligned with the broader target to decarbonize the US economy in the same timeframe. One critical aspect in reaching these goals is that all federal procurement needs to produce net zero emissions by 2050 and a "Buy Clean" policy that endorses the use of lower emissions construction materials. Another goal is for the government to use 100% CO_2-free electricity by 2030, and there is also a commitment by the USA to lower net GHG emissions by 50–52% under 2005 levels by 2030. Reaching these goals would demand some 10 GW of additional clean electricity production by 2030. In general, the specific goals by the US government are very demanding, and if successful, these policies will drive substantial emissions cuts as the federal government is the largest energy user and landowner in the USA. To elaborate this a bit more, the US Department of Defense (DOD) is one of the major consumers of energy in the world (the largest institutional consumer of petroleum) and is responsible for approximately 80% of all US government energy consumption. Strategically, the USA has aligned these targets with a positive goal of creating new jobs and improving infrastructure. For example, all federal vehicle purchases will be carbon neutral by 2035; additionally, all light-duty vehicle purchases need to be zero carbon by 2027. The implementation of these executive orders will therefore also strengthen the domestic supply chains. These targets extend to building infrastructure as well, and therefore, the US government's buildings portfolio is mandated to be carbon neutral by 2045 with a shorter-term goal of 50% GHG emissions reduction by 2032. The Biden administration has also stated a goal to allocate 40% of these climate investments to communities that are disadvantaged and have suffered from disproportionate climate change impact. As part of the general approach to these policies, the state level and corporations are given the freedom to make their own goals including structuring their renewable energy portfolio standards.

Key Climate change policies in the USA (Whitehouse 2023).

- Reducing US greenhouse gas emissions 50–52% below 2005 levels in 2030
- Reaching 100% carbon pollution-free electricity by 2035
- Achieving a net zero emissions economy by 2050
- Delivering 40% of the benefits from federal investments in climate and clean energy to disadvantaged communities.

When the European Union (EU) formulated its Nationally Determined Contributions (NDCs) under the Paris Agreement in 2015, it pledged to reduce greenhouse gas (GHG) emissions by at least 40 percent by 2030, relative to 1990 levels. In December 2019, the European Commission made new strives and published the European Green Deal, stating new targets and approaches in achieving climate neutrality. The new agreement was made in spring 2021 by creating a Climate Change Law, making the goal of becoming carbon neutral by 2050 and the new stricter 55% GHG reduction target legally binding. The negotiation on the proposed specific directives was initiated in the European Parliament, European Commission, and member states. In June 2021, the Council officially passed the European climate law, a significant component of the European Green Deal. This law enforces the legal obligation for EU member states to achieve both the 2030 and 2050 climate objectives. The progress of the EU in reaching the climate policy goals has historically been somewhat positive. On March 8–9, 2007, the EU decided to adopt new environmental targets that were stricter than the Kyoto Protocol. The policy commitment was to reduce GHG emissions by 20%, to increase energy efficiency for saving 20% of EU energy consumption, to reach the point of 20% of renewable energy use and finally to reach 10% in the share of biofuels in fuel consumption by 2020 (from 1990 level). Impressively EU was able to reach these targets; however, only 21 of 27 countries were able to reach their individual targets. Additionally, in 2020, EU was also able to reach the goal of 20% renewable energy in their energy portfolio and a 20% reduction of emissions in transportation.

Key Climate change policies in the EU (European Council 2023)

- A climate-neutral EU by 2050
- At least 55% fewer emissions by 2030
- From climate goals to EU law
- Financing the EU's climate transition
- Shaping global action.

Notably, *China* did not withdraw support of the Paris Agreement when President Donald Trump (June 1, 2017) announced that the USA would not continue to participate in the 2015 Paris Agreement. China has truly taken the effort and acted expeditiously during the past decade in transitioning to a carbon neutral and cleaner economy. It has captured the opportunities in several technological fields as a supplier of climate technology. China has also become the largest investor in renewable energy. It has reportedly invested twice as much (nearly $760 billion between 2010 and 2019) in renewable energy as the USA, also leaving EU in the second place. Due to these investments and the rapid pace of development in China, the country has emerged as a global leader in the total installed capacity of wind and solar energy. At the end of 2020, China had already 288 GW of wind energy installed and 253 GW of solar energy installed. Additionally, 30% of the wind turbine manufacturing is in China, and more than 70% of the solar photovoltaics are also manufactured there. Interestingly in terms of the EV market, China controls the global supply chain of key components manufacturing.

China climate policy has continued to follow the NDC targets and has also been updated in 2020. The new targets and measures state that China aims at reaching their peak in CO_2 emissions prior to 2030. In addition, they will be committed to achieve carbon neutrality by 2060. They are starting to reduce overall carbon intensity by over 65 percent before 2030 (compared to the 2005 level) and are targeting at growing non-fossil fuel consumption in primary energy to a level of 25% by 2030. They are also drastically increasing forest stock (plantations) volumes by 6 billion cubic meters by the year 2030 compared to 2005 level. By 2030, there are plans to significantly expand the installed base of wind and solar power capacity, aiming to reach a total capacity of over 1.2 billion kW. China has also been active in several areas of climate action and is successfully incorporating it to the national economic and social development plans. Although there is a hard road ahead in terms of industrial energy portfolio development and energy efficiency improvements, there has been a major commitment in GHG emissions mitigation in key industries.

Key climate change policies in China

- *Peak CO_2 emissions before 2030*
- Achieve carbon neutrality before 2060.

3.4 Greenhouse Gas Protocol and Science-Based Targets

The Greenhouse Gas Protocol (GHGP) is important to understand as we continue to discuss the impact of various human activities in GHG emissions. It was developed in the latter part of the 1990s in partnership between World Resources Institute (WRI) and the World Business Council for Sustainable Development (WBCSD) and as such is considered as the standard framework used for measuring and reporting GHG emissions and categorizing them into Scopes 1, 2 and 3 based on the source.

The GHGP includes a set of standards that are used for many sustainability reporting, managing, and policy activities both in public and private sectors. It encompasses also relevant value chains and mitigation actions. The reporting protocol is the most widely used of its kind globally and several other reporting standards such as Global Reporting Initiative (GRI) and International Integrated Reporting Council (IIRC) are using it in their reporting standards. GHG protocol has developed many methodologies and calculation tools for assisting in calculating GHG emissions of their operations and for the evaluation and planning of mitigation projects.

According to the GHGP, the GHG emissions can be categorized into three different scopes. *The Scope 1* is defined to include emissions that come from company controlled or owned sources. This means any emissions from the company's own operations are included. There are three additional categories: *emissions that are released in industrial processes* (e.g., process heat, cement manufacturing, chemicals), *building onsite energy use* (e.g., heating and cooling) and related sources for industrial heating (all GHG emitting fuels) and refrigerants, and *company vehicles emissions* (owned and leased). *The Scope 2* emissions are defined as the indirect emissions that come from the generation of purchased energy and therefore related to consumed energy that is provided by energy companies or utility providers. This includes, for example, electricity, heat, steam, and cooling. Technology development can influence these categorizations. For example, when companies are moving towards using electric vehicles (EVs) in the operations, this shifts the company vehicles emissions from Scope 1 to Scope 2.

The Scope 3 emissions encompass indirect emissions that are not included in Scope 2 and that come from the entire value chain of the company. A comprehensive approach to GHG accounting involves adopting both a value chain and a life cycle perspective. The Scope 3 approach follows a "cradle-to-grave" principle, considering emissions throughout the entire life cycle of a product or process. This includes accounting for emissions associated with raw materials extraction, manufacturing, transportation, storage and warehousing, sales, end-use, and disposal stages. Examples of Scope 3 include *upstream operations* such as business travel and employee commute, waste generations, life cycle emissions of all goods and services purchased, and all transportation and distribution activities by suppliers and customers. Additionally, any fuel and energy-related activities that fall outside of Scopes 1 and 2 are considered. Lastly, in terms of capital goods companies, they are expected to incorporate the evaluated total "cradle-to-gate" emissions of capital goods in the year of acquisition (GHG protocol). For *downstream activities,* it may include for example leased assets to other companies or organizations, emissions that are coming from the utilization of products sold, and finally the assessment on how their products are disposed of or recycled. The Scope 3 can be quite challenging task for many companies; however, diligent use of it helps companies lower their GHG emissions as a significant part of any company's carbon footprint comes from their value chain.

Again, the Scopes 1 and 2 are considered mandatory reporting, and Scope 3 is voluntary and as such demands expertise and company investments to address and monitor properly. It can be stated that companies that are successful in reporting Scopes 1–3 do have the opportunity to gain a competitive advantage in sustainability and differentiate themselves in their respectful value chains.

Carbon neutrality as a general concept means that companies need to balance their carbon emissions by either preventing or removing them from their own operations or by purchasing the equivalent amount of carbon offsets. Due to the ambiguity, consistency, and lack of standards, the *Science-Based Targets Initiative (SBTi)* was launched in October of 2021 with a purpose of defining and clarifying the formulation of net zero carbon strategies. This initiative is a collaborative effort between Carbon Disclosure Project (CDP), United Nations Global Compact, World Resources Initiative (WRI), and the Worldwide Fund for Nature (WWF). The basic approach is to measure the Scope 1–3 emissions, create a management plan for carbon that includes Greenhouse Gas (GHG) reduction targets, and use a precise methodology and tracking and annual updates. SBTi is purposed to give corporations tools and processes to reduce their GHG emissions and to make sure that the targets are aligned with the Paris Agreement. Using this process gives credibility to the companies using it. Most companies are setting their goals using net zero a two-step process: first reduction milestones by around 2030 and ultimately net zero milestone by 2050. Many of the government goals around the world have been formulated similarly and thus follow the same structure and process.

These corporate and governmental policies and voluntary agreements, specifically related to energy industry's efficiency improvements and overall reduction of GHGs are not new and can be traced back 30 years. Admittedly, the latest carbon zero targets are an adaptation and solidification of industrialized countries' energy industry's plans and extensions to include developing and emerging economies. These developments can be categorized to include solely voluntary plans and targets, proactive activities that are purposed to adjust and respond to future mandatory targets or incentives (e.g., regulation, carbon tax policy) and programs that are purposed to comply with new mandatory regulations and policies. The SBTi created opportunities for finance and investment sector to, in a transparent and structured way, add the carbon zero targets to their investment portfolios and to achieve a strategic portfolio alignment towards carbon neutrality. The new approach will facilitate the validation of the selected climate targets with the up-to-date climate science. It also serves as a linkage between real economy, investing, and finance. The timing for implementation is ideal because climate change has been demonstrating significant risks for a while now. This has made it urgent and crucial to prioritize financial investments and shift global capital flows toward climate technology and sustainable energy.

What is then *EU Taxonomy*? The European Commission (EU) has for the past decade been driving a very strong line for its member countries to reduce GHG emissions and fight climate change. It recently developed and very rigorous EU Taxonomy process to allow companies to comply and justify them being categorized as compliant with the objectives for climate change mitigation. These rules are initially encompassing the energy, transportation, building and manufacturing sectors with nuclear and natural gas sectors still waiting for consideration on their role in the compliant economic activities. It is estimated that it now covers economic activities that are liable for about 80% of EU GHG emissions.

This approach is aimed at also impacting the investment activities towards low-carbon technology platforms by giving the investors tools and visibility for informed decisions. Interestingly both SBTi and EU taxonomy are similar in terms of using fact-based benchmarking when organizations or economic activities are evaluated and aligned with their goals and criteria. In terms of SBTi, the alignment is in Paris Agreement and the well-below 2 °C global temperature raise limitation. In the case of EU Taxonomy, in addition to Paris Agreement and GHG emissions, SDGs and other related aspects are considered. It should be mentioned that EU Taxonomy also uses ESG framework and UN Global Compact and OECD guidelines and therefore integrates well with the broader global principles in driving sustainability. More specifically, there are six environmental objectives in the EU Taxonomy: *1. climate change mitigation, 2. climate change adaption, 3. sustainable use and protection of water and marine resources, 4. transition to a circular economy, waste prevention and recycling, 5. pollution prevention and control, and 6. protection of healthy ecosystems.* Further, the economic activities that are part of the taxonomy are defined in a Statistical Classification of Economic Activities or Nomenclature statistique des activités économiques dans la Communauté européenne (NACE). Mainly the Scope 1 activities (emissions from own or controlled sources) are included with a few exceptions from Scope 2. Finally, the main purpose of the EU Taxonomy is to facilitate and guide EUs stakeholders to transition to a more sustainable economy. It provides reporting assistance for the Sustainable Finance Disclosure Regulation (SFDR) and the upcoming Corporate Sustainability Reporting Directive (CSRD), adopted in April 2021 by the European Commission. This will amend the reporting requirements of the earlier Non-Financial Reporting Directive (NFRD). For clarity, the EU Taxonomy is by no means a list of economic activities that are only allowed by the investors in EU, nor does it create regulated targets for environmental performance including climate change mitigation. However, reporting is mandatory using EU Taxonomy for all large companies (financial and non-financial) that are also subject to publishing non-financial information.

3.5 Corporate Policies and Targets

Carbon neutrality as a general concept means that companies need to balance their carbon emissions by either avoiding or removing emissions from their operations, products, and associated value chains or by purchasing the equivalent amount of carbon offsets. As earlier discussed, due to the ambiguity, consistency, and lack of standards in setting up targets and following up on them, the SBTi was launched in 2015 with a purpose of defining and clarifying the formulation of net zero carbon strategies. The basic approach is to measure the scope of 1–3 emissions, create a management plan for carbon that includes GHG reduction targets, use a precise methodology, and enable tracking and annual updates. Using this process gives much-needed transparency, accountability, and credibility to the companies with ambitious goals.

Most companies are setting their goals using a two-step process for net zero or decarbonization targets: emission reduction milestones by 2030 and ultimately net zero final targets by 2050. Many of the government goals around the world have been formulated similarly and thus companies follow the same structure and process. These policies and voluntary agreements are in many ways extensions to former plans and targets.

These developments include solely voluntary plans and targets, proactive activities that are purposed to adjust and respond to future mandatory targets or incentives (e.g., regulation, carbon tax policy) and programs that are designed to comply with new mandatory regulations and policies. The SBTi has created opportunities for finance and investment sector to add the carbon zero or decarbonization targets to their investment portfolios and to achieve a strategic portfolio alignment towards carbon neutrality. This new approach facilitates the validation of the selected climate targets with the up-to-date climate science, in a transparent and structured way. The timing is crucial, and the role of financial investments and the global redirection of capital inflows are rapidly shifting to a more sustainable energy infrastructure. The monetary needs are tremendous and estimates of the necessary additional investments to, for example, energy infrastructure for achieving Paris Agreement goals by mid-century range between $2 trillion and $4 trillion (IEA 2021). This means major commitments to the best available technologies for energy efficiency improvements, green energy technology, and infrastructure investments. In addition, energy companies and governments need to invest considerably to emerging green technology development.

To achieve the highly challenging carbon zero goals, the power utility companies need to generate electricity and heat emission-free. Especially in terms of electricity, the options and technologies are well known, including solar, wind, nuclear and hydropower. Necessary advancements are still needed in developing hydrogen economy concepts and more effective storage solutions for electricity or other forms of energy carriers. Smart and flexible grid and dynamic power source management for continuous availability is also needed in this context. Electrification in terms of energy-intensive equipment and systems is one part of the equation and also includes other energy forms for transportation such as hydrogen, methane, ammonia, and biofuels options. Energy efficiency remains one of the key areas for continuous improvement. It has been estimated that digital technology, including smart energy systems and big data, can cut global emissions by as much as 20% (Royal Society 2020). This is because more intelligent use of energy creates major savings and reduces idle misuse. It is also crucial that investments must focus not only production but also in consumption reductions.

Companies have also been investigating various CO_2 removal technologies from point sources or from the atmosphere. The most typical technology platforms include Carbon Capture and Storage (CCS) and Carbon Capture, Utilization and Storage (CCUS), in which the CO_2 is also stored permanently into oil and gas reservoirs, or deep saline reservoirs. The reuse includes pressurization of CO_2 into a liquid form and its transportation to sites of use. There are several applications for captured CO_2 including enhanced oil recovery (90% of current CO_2 use); CO_2 can be permanently locked into cement (even with 25% of CO_2 by weight), which also reduces the use of lime (major source for CO_2 emissions). The CO_2 can be combined with hydrogen in the process of creating synthetic diesel, jet fuel and gasoline. Especially, aviation industry sees this approach as an attractive alternative to reduce emissions.

Many of these CO_2 use cases are still emerging, and thus, the government policy and tax credit support are critical to move this new market forward. Current incentive, in the USA, is \$35 per ton of using CO_2 for products and \$50 per ton for geological permanent storage (IEA 2018). In European Union (EU), Finland was the pioneering country to introduce carbon tax. The prices currently range depending on the EU country between \$0.08 and \$137.00 per ton CO_2-eq (Tax Foundation 2021).

Progressively, more asset owners attend to climate change as a principal priority. This is a major challenge as the material risks related to the company assets and its supply chain need to the factored in. Also, transitional costs from capital investment and policy changes that potentially alter customer and consumers behavior are imminent, affecting asset valuations and the investment portfolios. Especially, the Scope 3 emissions and the related risks can be hard to estimate.

There is also a shortage of accurate data, and double counting is still an issue and therefore unclarity exists especially on who is responsible for reported emissions. Regardless of the developments in carbon zero target setting, its transparency and reporting structure, the world (and investment companies) sees a need for traditional fossil-based energy. This applies especially to the developing countries. It has been reported that the 60 largest banks and investment companies globally have invested \$3.8 trillion into fossil fuels since Paris Agreement (Bernardelli et al. 2022). In addition, commercial lenders have financed coal-based power and utilities in total \$1.5 trillion globally since 2019 (IEA 2021). The global commitment of reaching the 1.5 °C target is a Herculean task, and regardless of all the reported financial investments, we are considerably behind our target. The International Energy Agency (IEA) has stated that unless coal is soon obsolete and not part of the investment companies' portfolios, we will not reach the Paris Agreement goals.

This remains the case despite the growing vocal commitment of global business leaders, the investment community, and policymakers to energy transition. The research indicates that the world continues its trajectory to increase fossil fuel dependency. For example, there are currently nearly 200 coal-fired power plants being constructed in Asia, with China leading the way with 95 plants under construction, followed by India with 28 in construction and 23 in preconstruction phase (Littlecot et al. 2021; Shearer 2016). The fact that these plants will be operational for at least 30 years makes the 2050 targets for CO_2 emission reduction highly challenging especially when there are more than 1000 operational coal plants in China, and 145 are in planning phase (Cui et al. 2021).

Some activities to expedite the phasing out of coal have been initiated, for example, as part of the Energy Transition Mechanism (ETM), the Asian Development Bank (ADB) is identifying coal power plants to buy and retire across Southeast Asia. They are also converting them to biomass-based energy production by using specific funding to support such investments (ADB, 2022). The growth of coal-based energy production in 2021 was projected to be 9% (Coal 2021). It should also be stated that it is estimated that over 35% of global energy production is from using coal, 25%, comes from natural gas, 16% from hydropower and 10% from nuclear. The renewables including bioenergy, solar and wind account for the remaining 12% of total energy (IEA 2021).

A brief analysis from various public reports and websites of large corporations operating in several industries that are large contributors to climate change shows evidence of various ambition states with both clearly defined and general targets. In addition, the "how?" is mainly focused on energy and statements on SBTi-related plans. The evidence found on concrete actions displays relatively small initiatives and indications of longer implementation time span than targets indicate.

During the process of writing this book, several interviews were conducted with large global corporations, including in these industry sectors. The executives interviewed had similar viewpoints to the challenges in corporate policy and target setting. The concerns were mainly on the high-level target setting with a lack of governance and management system transparency and dialogue. Also, lack of middle management and experts' involvement in preparing actionable policies and realistic plans and targets was raised. The role of corporate-funded initiatives versus business unit-level operational and manufacturing technology-related investments was not clear.

This may potentially slow down some of the investments needed and create lack of commitment and accountability in reaching the highly ambitions carbon neutrality goals in these companies. In addition, a questionnaire was constructed with 100 participants from various industries, including different levels of management (senior, middle management, and operations). In total, 21 questions were used, and only those individuals with relevant roles in decision making related to climate change strategies were included. Some of the questionnaire results on policy and its implementation are highlighted as follows:

Selected question #3: Does your company have a carbon neutrality target and if so, by what year?

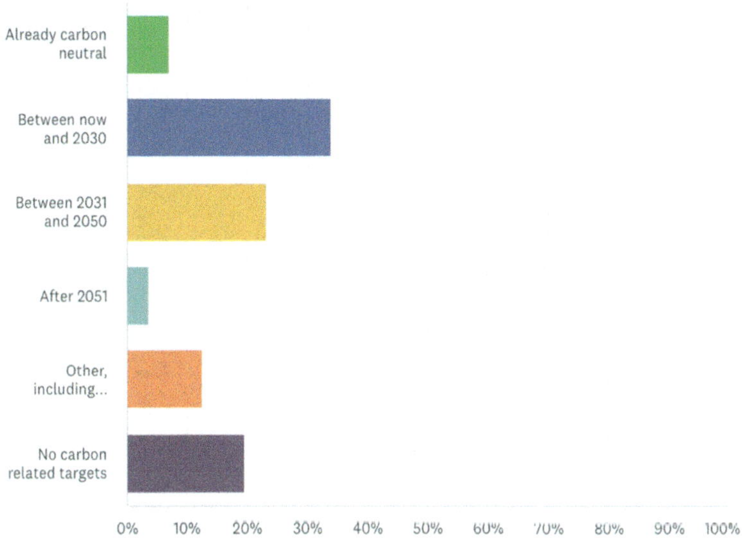

The participating companies' goals were for 2030 or 2050. "Other, category included carbon emissions reduction goals but not "neutrality," which is selected by companies that only have benchmark goals at this time.

Selected question #4: How important do you feel carbon neutrality commitments are to the future competitiveness of your company?

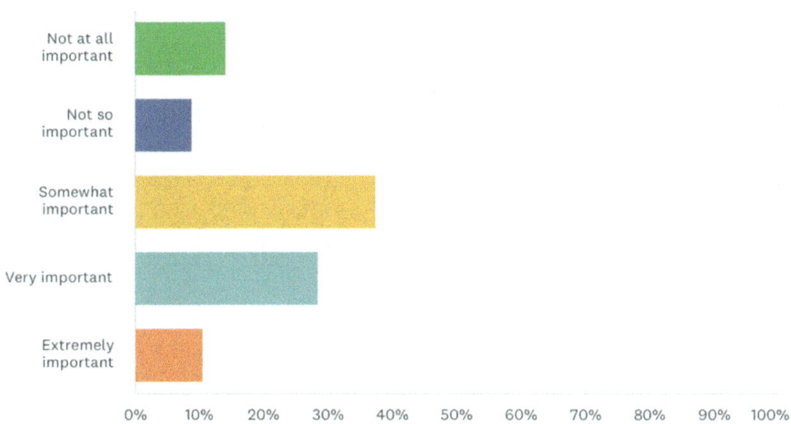

"Very important" or "extremely important" was selected by 22 respondents. "Not so important" and "not at all important" were selected 13. There were eight respondents who elected, "not at all important," three of them did not have carbon neutrality goals, and one was already carbon neutral.

Selected question #11: How are investments towards carbon neutrality funded?

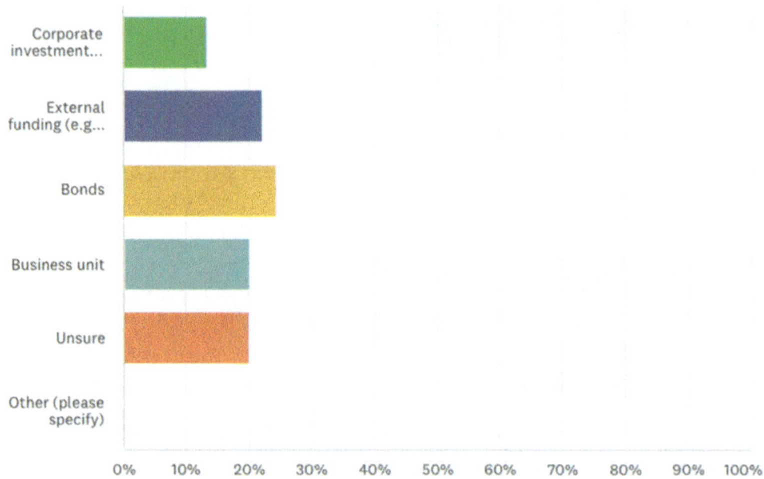

The most common answer was using "bonds," which was followed by "external funding." The third most popular answer was, "business unit," followed by "unsure." The least selected option was "corporate investment budget." This result was aligned with the interviews performed and thus reinforced the observation of lack of clear ownership to the manufacturing technology investments and funding responsibility.

Selected question #15: Carbon–neutral targets are typically made at the corporate level and include stakeholder involvement. How involved were the operations and business units in setting up the targets?

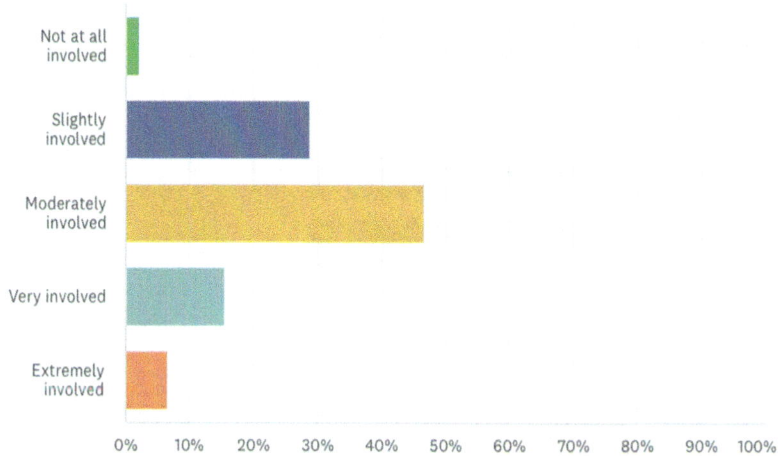

The results indicated that there is "moderate" or "slight" involvement. However, also "very involved" and "extremely involved" were selected. This gives a somewhat positive outlook to the inclusive and involving climate change strategy development in corporations that were participating to the study.

Selected question #18: Select all of the following decarbonization strategies your company plans to or currently uses, if any, to reach your carbon neutrality goal?

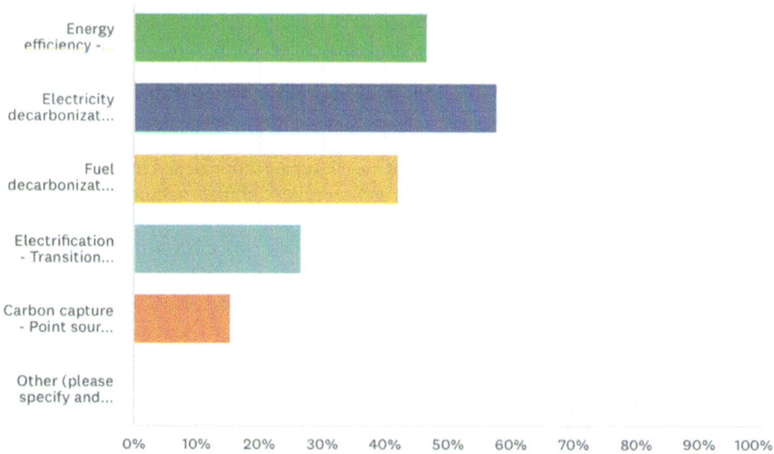

The outcome of this question shows the following ranking: (1) "electricity decarbonization," (2) "energy efficiency," (3) "fuel decarbonization," (4) "electrification," (5) "carbon capture," and no responses for "other." It should be noted that "electricity decarbonization" includes renewable energy.

Selected question #20: Please rank the following in the order you believe is most to least important in the path to decarbonization.

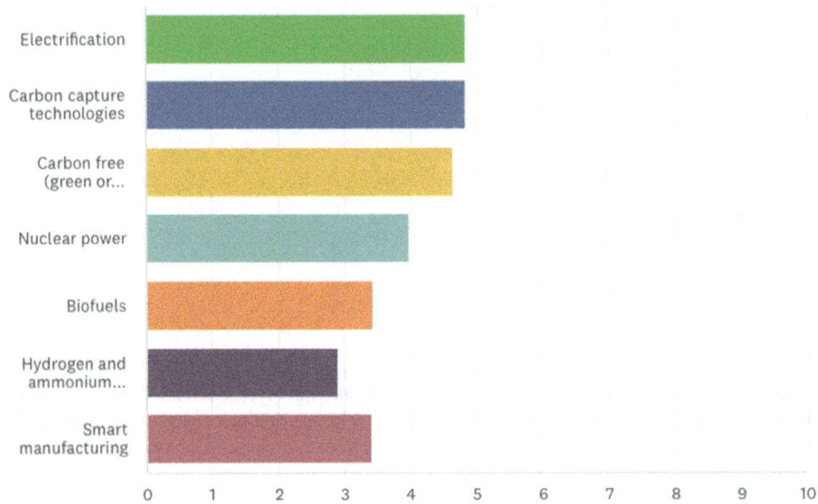

The positive view to carbon capture technologies was clear, and the responses were ranked as follows: (1) carbon capture technologies, (2) electrification, (3) carbon-free electricity, (4) nuclear power, (5) biofuels, (6) smart manufacturing, and (7) hydrogen and ammonium platforms. Interestingly, the role of nuclear power has become more important than the biofuels.

Although this questionnaire was limited in scope and participation, it did highlight many of the concerns regarding policy statements by corporations and governments. The road to implementation of the very ambitious targets demands systemic approach and is a long organizational learning process. It is clear that social, organizational, and human behavioral aspects need to be considered when progress and success of climate actions are evaluated.

References

ADB (2022) Asian Development Bank ISBN 978-92-9270-074-4. https://doi.org/10.22617/FLS230039

Bernardelli M, Korzeb Z, Niedziółka P (2022) Does fossil fuel financing affect banks' ESG ratings? Energies 15:1495. https://doi.org/10.3390/en15041495

Coal (2021) https://iea.blob.core.windows.net/assets/f1d724d4-a753-4336-9f6e-64679fa23bbf/Coal2021.pdf Global Cold Chain Report (2020) https://www.gcca.org/resources/2020-global-cold-chain-capacity-report

Cui RY, Hultman N, Cui D et al (2021) A plant-by-plant strategy for high-ambition coal power phaseout in China. Nat Commun 12:1468. https://doi.org/10.1038/s41467-021-21786-0

Environmental Policy, Cocklin C, Moon K (2020) International encyclopedia of human geography. 2nd edn. Elsevier. ISBN 9780081022962

European Council (2023) https://www.consilium.europa.eu/en/policies/climate-change/

IEA (2021) Net Zero by 2050. IEA, Paris. https://www.iea.org/reports/net-zero-by-2050

Kashani ES (2009) Policy gap in science and technology policy analyses: promoting innovation in conflict with technology risk. J Sci Technol Policy 2(1)

Knill C, Tosun J (2020) Public policy: a new introduction. Bloomsbury Publishing

Littlecott C, Roberts L, Şenlen Ö, Burton J, Joshi M, Shearer C, Ewen M (2021) No new coal by 2021: the collapse of the global coal pipeline. E3G. https://9tj4025ol53byww26jdkao0x-wpengine.netdna-ssl.com/wp-content/uploads/No-New-Coal-by-2021-the-collapse-of-the-global-pipeline.pdf

Shearer C, Ghio N, Myllyvirta L, Yu A, Nace T (2016) Boom and bust 2016. Sierra Club. https://www.sierraclub.org/sites/www.sierraclub.org/files/uploads-wysiwig/Final%20Boom%20and%20Bust%20report_0.pdf

Tax Foundation (2021) https://taxfoundation.org/carbon-taxes-in-europe-2021/

The Royal Society (2020) Digital technology and the planet: harnessing computing to achieve net zero. ISBN: 978-1-78252-501-1

Weekes WH (1981) Action learning—an open system approach to management education. Asia Pacific J Human Resources 18(3):9–13

Wergin JF (1976) The evaluation of organizational policy making: a political model. Rev Educ Res 46(1):75–115

Whitehouse (2023) Retrieved March 2023. https://www.whitehouse.gov/

Wies R (1994) Policy definition and classification: aspects, criteria and examples. In: Proceedings of the IFIP/IEEE international workshop on distributed systems: operation and management, pp 10–12

Chapter 4
Sustainable Business Models Driving Carbon Neutrality

Abstract In this section, we explore sustainable business models, with a particular focus on those revolutionizing the climate technology sector. We look into influential concepts shaping businesses' sustainability approaches that are critical for greener and more responsible future. One driving force is natural capitalism, which urges businesses to harness market forces to enhance resource efficiency, minimize waste, and foster innovation. Biomimicry, an inspiring concept, draws inspiration from nature's genius designs and industrial ecology provides a transformative perspective, viewing industrial processes as interconnected ecosystems. Cradle-to-cradle design presents a visionary paradigm, challenging businesses to create products and materials that are safe, fully recyclable, and biodegradable. Additionally, Life Cycle Assessment empowers businesses to evaluate their products' environmental impact comprehensively. We also embark on an exploration of the circular economy and bioeconomy, sustainable economic models that emphasize waste reduction and renewable resource utilization. Intertwined within these models are the United Nations Sustainable Development Goals, guiding businesses towards addressing global social, economic, and environmental challenges. Furthermore, we present a case example highlighting the textile industry's embrace of carbon zero business models. Through innovative strategies, this industry demonstrates its commitment to sustainability and a carbon-neutral future.

Keywords Sustainable business models · Biomimicry and industrial ecology · Cradle to cradle · Life cycle assessment · Circular economy · Bioeconomy · Sustainable development goals

4.1 Background

A business model can be considered as the core strategy of any business that enables it to do be profitable in short and long term. In general, the core description of a business model includes a reasoning on how the company or organization will create, deliver, and capture value. It includes feasibility on how the company is able

M. Hakovirta, *Carbon Neutrality*, Springer Climate,
https://doi.org/10.1007/978-3-031-45202-4_4

to deliver its products and services, create attractiveness to its offering and finally establish viability for the company to be profitable. The feasibility for a company to deliver its value proposition demands key partners, activities, and resources. Partners are crucial in optimizing operations, reducing any risks, and developing or gaining a competitive advantage. Identifying and designing the most important activities for any company gives focus to resource allocation and makes sure that company's value proposition is been built and continuously developed. It also makes sure that the right resources are available and are aligned with the activities. The ability to attract customers demands a well-positioned selling proposition that differentiates from the competitors. All products and services sold need to be well described, communicated, and clearly show how they meet the customer needs. Understanding customers is critical in both identifying them by segments (e.g., location, age, gender, interests) and in building appropriate relationships with them. This all needs to be well managed, and the appropriate way to interact with them is important. Different channels for delivering company's value proposition to targeted customers must be identified. These can be company's own channels or based on using partners or both. In order for a company to be able to operate and remain profitable, it has to have a cost structure that is affordable. This includes costs from activities such as R&D, marketing and sales, infrastructure investments, and so on. The revenue streams that are coming from various sources needs to be able to bring enough earnings based on the cost structure.

The above-described concept of business model is generally well accepted by practitioners and scholars (Osterwalder and Pigneur 2010). However, when you add sustainability in its various forms into the same conceptual model, one has to look at optimization and balancing acts of short- and long-term profitability, the value proposition, and its alignment with the customer segmentation and who are the chosen partners and stakeholder that are committed to sustainability. In theory all the components of a business model can be adjusted to sustainability with an adjusted mindset. There are several business model concepts that we will introduce in this chapter that capture some unique characteristics of sustainable business models. Most important is to understand the life cycle thinking and the optimization and balancing of the set of activities that defines company's ability to differentiate and add value. This needs to be done in order to create a true sustainable business model, regardless of the technology, product or service offering.

4.2 Natural Capitalism

The Natural Capital is a concept that has evolved during the 1990s and was described and further refined by authors of "Creating the Next Industrial Revolution" Amory Lovins, Hunter Lovins and Paul Hawken in 1999 (Hawken et al. 2013). The principal idea of Natural Capitalism is that the economy is reliant on natural resources and ecosystem services provided by nature. It can be stated that the concept itself criticizes the current economic system or traditional industrial-based capitalism. The basic idea of Natural Capitalism is that all raw materials and commodities such as air, water, minerals, soil, fossil fuels, and forests have been stored by nature for a few billion years. Now humankind is using this precious capital and for its own benefit for building industrial capitalism.

From a systems perspective, many of these systems are responsible for supplying basic things we all need including regeneration of atmosphere and water purification. The conceptual framework of Natural Capitalism deals with the interdependency between the manufacturing production, the use of human capital, and the maintenance and supply of natural capital. Ecosystem services are on the other hand a concept that defines the set of ecosystem functions that are affecting human well-being. These functions are the natural capital that is provided to humans by the natural environment and ecosystems. Examples of these ecosystems include agricultural ecosystems, forest ecosystem, and aquatic ecosystems. It has been estimated that the entire earth's total ecosystem services monetary value is as high as $33 trillion per year.

In principle any economy demands four types of capital: 1. human capital, including labor, skills and competencies, and organization, 2. financial capital, including cash and monetary investments, 3. manufacturing capital, that is appropriate infrastructure, machinery, process engineering and manufacturing sites, and finally natural capital. Looking at the concept of Natural Capitalism, the first business model-related strategic approach is to examine the opportunities to improve the productivity of natural resources. This means to produce more from less or to be more efficient in using resources and by applying lesser polluting processes. Another aspect of this is to increasingly add closed-loop production systems that are waste fee and do not create any toxicity. This requires a redesign of industrial systems by using principles of nature. Another aspect of this is to business model changes with for example circular economy approaches and with more focus on services than physical products. Circular economy model incorporates different ideas that are solution-based and include remanufacturing and recycling. Natural Capitalism also includes major reinvestments in to restore, preserve, and sustain our planet's ecosystem. At its very best the Natural Capitalism looks for a win-win situation with a balanced approach in which companies will be looking for creating a sustainable competitive advantage (Lovins et al. 1999).

4.3 Biomimicry and Industrial Ecology

Biomimicry is an approach to imitate nature and life in order to create innovation that is looking for sustainable solutions. It uses ideas and principles of natural selection-driven solutions adopted by nature and translates these principles to engineering. The concept of Biomimicry dates back to 1950s when Otto Schmitt Otto published his work attempting to produce a physical device that mimicked the electrical action of a nerve. In 1960s, Jack. E. Steele used a term, bionic, and defined it as systems that are copied from nature. Finally, Benyus publicized the general conceptual idea in her book on Biomimicry in 1997 (Benvus 1997). According to the Biomimicry, the natural selection and evolution that took place for over 4 billion years eventually has produced the solutions that our societies are founded on. A good example of this is that all living organisms were evolved and eventually developed their composition and structure over long geological timeframe of the existence of planet earth. The same way Biomimetics is the enabler for novel technologies that are often founded or inspired by existing biological solutions at macro and nanoscales. Many of the engineering achievement are in actuality based on nature's through time solved engineering challenges including self-assembly, self-healing, resistance to environmental exposure such as UV-light and hydrophobicity and harnessing solar energy. These and many other solutions are only now scientifically comprehended and fundamentally understood including having developed the ability to use these nature made engineering solutions to solve some of the global engineering challenges.

To conclude Biomimicry can be stated to promotes the understanding of nature and its design and defines that association as 1. Nature is the model for sustainable design and processes, 2. Nature is the measure for successful solutions, and 3. Nature is our mentor. (Hargroves and Smith 2006). The current scientific and engineering solutions include many examples of Biomimicry at work. The most classic example is the lotus effect that is used in many of the today's materials engineering application. The concept is to use the self-cleaning properties that the leaves of Nelumbo, the lotus flower has. These leaves have micro and nanoscopic architectural morphology on the surface. The effect minimizes water droplet's ability to wet the surface. The same concept is used in many self-cleaning and hydrophobic surfaces such as optical lenses, solar panels, frying pans, and industrial machinery.

Industrial Ecology studies industrial systems and looks at optimizing them by aiming at identifying and developing different approaches and solutions to reduce the environmental impact. The field is fairly new and is focused on studying the industrial systems and how such systems (manufacturing, energy plants and extraction of natural resources) work. The impetus is to create novel solutions for approaches to use less resources and also by finding new uses for industrial waste and by-products. These industrial approaches and processes also include various value chain activities that convert these resources to products and services that are targeted at meeting the various demands of the society. As a concept Industrial Ecology has been referred to be the "the science of sustainability." It has a primary goal of promoting sustainable development at the local, regional, national, and global level. And it studies the influences of economic, political, regulatory, and social factors, their flow, use, and transformation of resources (Ehrenfeld 2004).

4.4 Cradle-to-Cradle and Life Cycle Assessment

The principles of life cycle thinking include both cradle-to-grave and cradle-to-cradle look at the entire life cycle of a product. When the product is wasted at the end of life, the conceptual model is cradle-to-grave. On the other hand, if the product at the end of life is reused or recycled the model is called cradle-to-cradle. This thinking and the methodology to make such assessments deliberate the impacts to GHG and other emissions at each stage of a product's life cycle. The conceptual model starts from the extraction of the used natural resources and extends to include each stage of manufacturing, transportation, use of product, disposal, recycling, and reuse.

The cradle-to-cradle concept proposed by William McDonough and Michael Braungart in 2002 suggests a new approach to designing human systems that can solve the natural conflicts between economic growth and environmental health. This concept aims to replace the unsustainable "cradle-to-grave" approach with a more sustainable one that considers the value of life and the sustainability of future generations (McDonough and Braungart 2010).

A key aspect of the cradle-to-cradle model is the categorization of materials used in industrial processes into "technical" and "biological" nutrients. Technical nutrients are made of stable materials that are either inorganic or synthetic and can be used many times in continuous cycles without any loss in quality. These materials are designed to be retrieved and reused within sustainable manufacturing, and they are limited to non-toxic and harmless-to-environment materials. On the other hand, biological nutrients are materials or products that, after their use, can be disposed of in nature. They decompose and provide food for other life forms, and they do not negatively impact the natural environment. These materials are designed to be returned to nature and to be consumed by microorganisms in soil and other animals. The cradle-to-cradle framework proposes a sustainable manufacturing process that promotes the use of non-toxic and recyclable materials that can be endlessly reused in continuous cycles. This approach can potentially reduce waste and pollution, conserve resources, and promote the well-being of both the environment and the future generations. The three basic principles of the cradle-to-cradle concept are:

Waste equals food: In nature, waste from one organism becomes food for another. Similarly, the cradle-to-cradle model aims to eliminate the concept of waste by designing products and systems that can be continuously reused or recycled, creating a closed-loop cycle of materials.

Use of current solar income: Just like living systems rely on solar energy for sustenance, the cradle-to-cradle concept aims to use renewable energy sources like solar and wind for society's benefit, and for reducing reliance on non-renewable resources like fossil fuels.

Celebrate diversity in everything: Natural systems thrive on diversity and complexity, and the cradle-to-cradle framework encourages the use of a variety of materials, processes, and approaches to create resilient and adaptable systems that can respond to changing conditions.

Life Cycle Assessment (LCA) is a systematic approach used to evaluate the environmental impact of a product, service, or manufacturing process throughout its entire life cycle. The assessment encompasses all stages of the life cycle, starting from the extraction of raw materials from the environment and their processing (known as cradle), through to the manufacturing of the product (gate), its distribution and use, and ultimately, its disposal or end-of-life stage (grave). The International Organization for Standardization (ISO) has developed ISO14040 and ISO14044 standards, which provide a well-defined and standardized framework for conducting LCAs (Hauschild 2018).

The methodology of Life Cycle Assessment (LCA) involves four distinct phases: *Goal and Scope definition, Inventory analysis, Impact assessment, and Interpretation.* During the Goal and Scope definition phase, the analysis' primary objectives are determined, such as the assessment of product design alternatives, comparison of different products, or a specific impact in the product's life cycle. The project scope is also defined during this phase to ensure that the project goals are met. This phase is considered crucial in LCA work.

The *Inventory analysis* phase involves analyzing the environmental inputs and outputs associated with a product or service, including raw materials, energy usage, emissions to air and water, waste, and co-products. This phase is considered a core part of LCA, as it involves collecting and compiling the data required to study the system.

The *Life Cycle Impact assessment phase (LCIA)* investigates and evaluates the potential environmental impact of the system throughout its life cycle. The data is quantified into environmental indicators that are more intuitive and meaningful for comparison, interpretation, and business decision making.

Lastly, the *Interpretation phase* involves verifying and validating the conclusions drawn from the previous phases. The ISO14044 methodology provides various checks to test whether the conclusions are supported by the data. The LCA methodology is known by its attention to details and reliance on data, which distinguishes it from other frameworks that rely more on qualitative and narrative approaches. As a result, companies often choose to use LCA for product development, design, marketing, and sales, for more quantitative data-based reporting. Nonetheless, the LCA has its limitations, as its outcomes and conclusions are heavily influenced by the definition of the Goal and Scope, and, as with any methodology, the quality of the data used can impact the results.

4.5 Circular Economy and Bioeconomy

The *circular economy* is a novel approach to sustainable business that aims to generate value and prosperity by prolonging the lifespan of products, minimizing waste, and using resources in a more circular and efficient manner. This is accomplished by implementing enhanced design and service practices, as well as through the strategic utilization of waste, which is shifted from the end of the supply chain to the beginning. The circular systems that underpin the circular economy also encourage the reuse, sharing, repair, and refurbishment of products, with the ultimate goal of achieving closed-loop systems that reduce the need for new resource inputs. Extending the lifespan of products and improving manufacturing processes are key aspects of circular economy. The circular economy encompasses multiple definitions, one of which is provided by James Gaffrev. According to Gaffrev, the circular economy refers to the sustainable and sequential transformation of biological residues into bio-based products that can be shared, reused, remanufactured, recycled, or safely returned to the biosphere through organic and nutrient cycles. This definition highlights the importance of maximizing the value and utility of biological resources throughout their lifecycle, while minimizing waste and environmental impact. Non-profit organizations have played a vital role in educating people about the principles of the circular economy, which can be quite complex and are often mistaken as merely recycling-related.

Circular economy and *bioeconomy* are both two important concepts in sustainable business. The circular economy focuses on creating value and prosperity by eliminating waste and using resources in a more circular way, while *the bioeconomy uses renewable natural resources* as the raw materials for the manufacturing of materials, chemicals, and energy production. Both concepts are essential for achieving a more sustainable future and promoting a circular and regenerative economy. It's also important to note that while the circular economy is more focused on the use of materials and resources, the bioeconomy is focused on the use of biological or bio-based resources, such as biomass from forests and agriculture. Both of these principles of sustainable business hold the potential to generate economic, environmental, and social advantages. However, realizing these benefits requires collaboration among diverse sectors and stakeholders. More specifically, the circular economy is a business model that emphasizes sustainability and resource efficiency. In this model, companies focus on environmentally friendly product design and consider factors such as low-carbon footprint, reuse, and recycling at the early stages of product development. Materials selection involves a preference for recycled and renewable materials as inputs to manufacturing. Sustainable manufacturing evaluates carbon footprint, energy, water, and waste efficiencies usingLife Cycle Assessment tools. Supply chain management plays a critical role in sustainable business, with a focus on greenhouse gas accounting and sustainable transportation. Consumer behavior is also a crucial aspect of the circular economy, as promoting acceptance and active use of recycling can have a significant impact. Government policies on collection, reuse, and recycling of products are important, as are awareness and cultural factors. Recycling and reuse require new technologies and industrial standards, and policies and legislation need to address recycling opportunities and circularity. The bioeconomy on the other hand is a relatively new business concept that involves a shift from a traditional fossil-based economy to an alternative that uses renewable natural resources such as water, solar, wind, and biomass as the basis for materials, chemicals, and energy. It is seen as the new industrial evolution and is believed to be the next wave in world economic cycles according to Kondratieff's theory. By 2030, the economic gains associated with this development are estimated to be as much as 2.7% of the GDP in OECD countries. Biomass from forests and agriculture is the one of the most promising source of feedstock for the bioeconomy. The bioeconomy industry utilizes biotechnology and other engineering processes to produce sustainable goods, services, and energy. It plays a vital role in offering new approaches for sustainability and environmental well-being, which are critical for establishing a stable and socially responsible global society in the future.

4.6 United Nations Sustainable Development Goals

One of the most known and powerful frameworks in driving sustainability is the United Nations Sustainable Development Goals (SDGs) (UN DESA 2023) (Fig. 4.1). They were established in 2012 during the United Nations Conference on Sustainable Development held in Rio de Janeiro, with the aim of defining a set of shared and comprehensive objectives that could address the global environmental, political, and economic challenges faced by governments and corporations worldwide. The SDGs comprise 17 interconnected goals that range from climate change mitigation to natural resource management and poverty alleviation. The SDGs are built on the previous Millennium Development Goals (MDGs), which were launched in 2000 to combat global poverty. The SDGs are a part of the UN Resolution, also known as the 2030 Agenda, and are intended to be achieved by 2030, with concrete targets, indicators, and metrics to measure progress and performance. While most targets are set for 2030, some do not have a fixed deadline. The SDGs are vital for driving societal and business well-being, improving the quality of life for future generations, and addressing systemic sustainability challenges. The SDGs form a crucial part of the UN Resolution known as the 2030 Agenda and were primarily intended to be achieved by 2030. In 2017, the SDGs were revised to become more practical and attainable by incorporating specific targets, implementation indicators, and performance metrics for better assessment, comparison, and tracking of progress.

The SDGs encompass a large or broad extent or scope of critical issues that consist of climate change, natural resource conservation and efficiency, poverty reduction, and societal inclusiveness. Each 17 SDGs come with a specific objective. Goal 1 seeks to eradicate all forms of poverty by 2030. Goal 2 aims to eliminate hunger and malnutrition by the same year. Goal 3 aims to ensure good health and well-being. Goal 4 focuses on providing quality education for everyone and promotes continuous learning. Goal 5 of the United Nations SDGs aims to achieve gender equality and empower women. Goal 6 seeks to ensure universal access to clean water and sanitation. Goal 7 is centered around ensuring access to affordable, reliable, sustainable, and modern energy for all. Goal 8 promotes sustained, inclusive, and sustainable economic growth, full and productive employment, and decent work for all.

Fig. 4.1 UN sustainable development goals (SDGs)

Goal 9 focuses on building resilient infrastructure, promoting inclusive and sustainable industrialization, and fostering innovation. Goal 10 aims to reduce inequalities within and among countries based on various factors. Lastly, Goal 11 focuses on making cities and human settlements inclusive, safe, resilient, and sustainable. Goal 12 aims to ensure sustainable consumption and production, considering the need to improve the quality of living. Goal 13 targets urgent action to combat climate change and its effects. Goal 14 aims at conserving and sustainably using oceans, seas, and marine resources for sustainable development. Goal 15 is about protecting, restoring, and promoting the sustainable use of ecosystems, while also combating land degradation, and preserving biodiversity. Goal 16 targets at building peace and promoting inclusivity and a just societies, at every level. Finally, Goal 17 calls for partnerships to achieve these goals by improving the implementation and revitalizing the necessary partnerships for global sustainable development (Pradhan et al. 2017).

The SDGs comprise of 193 specific targets that outline sub-goals that are crucial for attaining the overarching goal. The UN closely monitors the progress of these goals through the UN Secretary-General's annual report on SDG progress (UN DESA 2023). Despite COVID-19 pandemic-related setbacks, the latest 2021 report reveals that progress towards achieving many of the SDGs by 2030 is not on track. Nonetheless, the world is paying attention, and COVID-19 may have served as a reminder of the consequences of human activity on global biodiversity.

The idea of sustainability has undergone changes over time and has been influenced by different factors that affect public awareness. However, the fundamental principles and frameworks are not entirely new to us. Sustainable business involves balancing short-term economic benefits with long-term benefits, while recognizing the value of natural resources as a shared resource to be respected and preserved. The desire to capture short-term gains is inherent in human nature and survival instincts and is often linked to cultural aspects. The concept of the "Tragedy of the Commons" refers to the tendency of individuals to consume a shared resource for their own benefit, without considering the impact on others. Balancing the competing demands of sustainable business is a challenging task, as companies face constant pressure to grow and sustain profitability, while also prioritizing long-term sustainability goals. Although there are examples of successful corporate leaders who have achieved this balance, doing the right thing is not always easy, as it may not satisfy all stakeholders. Optimizing sustainable practices requires a nuanced approach that considers the interdependencies of all stakeholders in the business, society, and the wider world.

4.7 Case Example of Industry Level Carbon Zero Business Models—Textile Industry

The textile and clothing industry is estimated to contribute around 10% of global CO_2 emissions. This industry has been grappling with substantial sustainability challenges for an extended period, and the projected rise in clothing consumption further intensifies the issue. The textile sector ranks among the most polluting industries globally, exerting a significant environmental impact, especially in terms of water and energy consumption, greenhouse gas emissions, and waste production. Moreover, the textile industry has long been associated with labor exploitation and human rights abuses, particularly in the developing countries in which workers typically are low paid, have long work hours, and often face bad working conditions.

To address these issues, the textile industry has a clear need to adopt a more sustainable and ethical approach. This requires a collective effort from all stakeholders, including businesses, consumers, policymakers, and civil society organizations. Textile companies have taken steps to reduce their environmental footprint, such as investing in sustainable materials, reducing water and energy consumption, and adopting circular business models that promote recycling and waste reduction. They work to improve labor conditions in their supply chains, by ensuring fair wages, safe working conditions, and freedom of association for workers.

Consumers also play an important role in promoting sustainability in the textile industry, by choosing to buy from brands that prioritize sustainable and ethical practices, by extending the life of their clothes by reuse, and by disposing of clothing responsibly using recycling or donations. Policymakers can support these efforts by introducing regulations and incentives that promote sustainable and ethical practices, such as carbon taxes, waste reduction targets, and labor standards enforcement.

In order to minimize the carbon footprint of textile production, several actions can be taken at every stage of the supply chain. For example, farmers can adopt sustainable agricultural practices carbon footprint reduction of natural fibers. Manufacturers can increase the use of renewable energy in their operations and adopt more efficient manufacturing processes that require less energy and generate less waste.

Designers can also have a major role in reducing the carbon footprint of textile production by designing clothes that are durable, easy to repair, and made from sustainable materials.

Consumers have the power to make a positive impact by making conscious choices when purchasing clothes. Opting for garments made from sustainable materials, such as organic cotton, recycled fibers, or innovative eco-friendly fabrics, can help reduce the environmental footprint of the textile industry. Additionally, consumers can actively participate in extending the lifespan of their clothes by practicing reuse and repair. This can involve donating or swapping clothes, mending minor damages, or repurposing items creatively. By adopting these behaviors, consumers actively support sustainable fashion practices and contribute to the reduction of waste and resource consumption.

End-of-life disposal or recycling of textiles is also critical for reducing the carbon footprint of the textile industry. Recycling or reusing textiles can significantly reduce greenhouse gas emissions compared to landfilling or incinerating them. The reduction of the carbon footprint in the textile industry will require new thinking and innovation throughout the value chain of the industry.

As for example cotton, although it is a natural fiber, it is far from environmentally friendly and can have a major negative burden on the environment. Cotton is typically grown in dry and warm regions, and it requires a considerable amount of water to grow. In some regions, such as India, inefficient water use in cotton production can result in a high-water footprint, with up to 20,000 liters of water required to produce just 1 kg of cotton. This can put a significant strain on water resources in these regions and contribute to water scarcity and environmental degradation. To address these issues, sustainable cotton production practices such as drip irrigation, crop rotation, and rainwater harvesting can be implemented to reduce water consumption and promote more sustainable cotton farming. Consumers can also help support sustainable cotton production by choosing to purchase clothes made from organic or sustainably sourced cotton.

The European Union is making textile waste recycling mandatory by 2025, as part of its efforts to create a more sustainable and circular economy. By developing new bio-based fibers from this waste, the textile industry can contribute to this goal and create innovative and environmentally friendly products. There are already some exciting developments in this area, such as using textile waste to create new fibers through mechanical, chemical, or biological recycling processes. These recycled fibers can be used to make a variety of products, from clothing and accessories to furniture and building materials. By investing in the research and development of these new materials, the companies reduce textile waste but also create new business opportunities and contribute to the carbon-neutral future.

According to the Ellen MacArthur Foundation, a global non-profit organization dedicated to promoting a circular economy, approximately 73% of the clothing material will end up in a landfill or incinerator. This means that the vast majority of clothing is not being reused or recycled and is instead contributing to waste and environmental pollution. Only a small fraction of clothing, approximately 12%, is currently being recycled, which highlights the urgent need for more sustainable practices in the textile industry. Currently, only 1% of the material used in garments is recycled into new clothes as the majority is downcycled to industrial use. This highlights the need for more innovative and sustainable recycling technologies that can enable us to transform used textiles into high-quality materials for new clothing.

It is also worth noting that the polyester used in many clothing items and identified as recycled is often sourced from recycled plastic bottles. While this approach is beneficial in reducing plastic waste and creating new products from existing materials, it is important to continue exploring and developing more sustainable and circular solutions for textile production and recycling. By investing in research and development and promoting sustainable practices throughout the textile industry, a more environmentally friendly circular economy-based approach can be achieved.

One of the major challenges facing garment-to-garment recycling is the issue of fiber quality. Fibers in textiles can become damaged during the normal process of wearing and washing, which can make them unsuitable for use in new garments. This means that it is not possible to tear apart a well-used garment simply mechanically and create new cotton fibers that are of high enough quality to be used in a new garment.

To address this challenge, new technologies are being developed that can help to preserve the quality of fibers during the recycling process. For example, chemical recycling methods can be used to break down the fabric into its constituent fibers without causing as much damage as mechanical recycling. Additionally, new materials are being developed that are designed to be more easily recycled and also biodegradable.

There are two main ways to recycle fibers in textiles: mechanical and chemical. The mechanical approach involves breaking down used textiles into their fibers and then mixing them with virgin fibers to make new fabrics. Chemical recycling, on the other hand, involves breaking down used textiles into their chemical building blocks and using those to create new fibers with properties that are equivalent to, or even better than, virgin fibers. This technique has the potential to create a closed-loop system for textile recycling, where used textiles can be transformed into high-quality textiles without the need for virgin materials. The chemical fiber-to-fiber recycling is challenging both technically and economically. This is because the resulting fiber cannot be more expensive than what the market will accept. However, there have been several pilot projects that have advanced such technologies and are now beginning to be introduced commercially.

In addition to fiber recycling, the collection and sorting of garments is a momentous challenge. The sorting is mainly done by hand and is slow, expensive, and not always accurate.

Automated systems using different technologies, such as infrared sensors and RFID tags, are being developed to improve the efficiency and accuracy of textile sorting. Infrared sensors can detect different types of materials based on their unique infrared signatures, and RFID tags can store information about the fabric type and other important data that can be used to sort garments automatically. However, these technologies are still in the testing phase and face several challenges, such as cost-effectiveness, scalability, and integration with existing recycling processes. It will require more research and development to optimize and commercialize these technologies for practical use in the textile recycling industry.

There have been various trials that have examined the possibility of chemically recycling cotton, which is responsible for almost one-third of all fibers utilized in textile manufacturing. The fundamental process stages of these trials entail the mechanical elimination of elements such as zippers and buttons, washing to extract dyes and other contaminants, dissolving cotton cellulose in a solvent, and then generating new fibers from the resulting pulp. Chemical cotton recycling offers several advantages over conventional recycling approaches as it can recycle blended cotton with other fibers and produce high-quality fibers that are comparable to virgin cotton. Nevertheless, there are some challenges associated with this technology include the use of expensive and hazardous solvents, concerns over the disposal of solvents, cost-effectiveness, and the need for significant investment to scale it up for industrial production. Despite these challenges, chemical cotton recycling has immense potential to decrease the environmental impact of textile production and disposal. One feature of the currently piloted approaches is that the fibers obtained from textile recycling may differ from one pilot to another. Some pilots have been able to generate familiar fibers through the recycling process, while others have developed new and innovative yarns.

Infinited Fiber, a spin-out from VTT Technical Research Centre of Finland, has recently announced that it has developed a unique cellulose carbonate fiber made from recycled cotton. In this process, the cellulose is modified with urea before being dissolved in a solvent. The application of heat breaks down the urea into isocyanic acid, which reacts with the hydroxyl groups in the cellulose to create carbamates. The resulting pulp is then regenerated into fibers using the viscose process, and old viscose plants can be utilized for this purpose. One of the major benefits of this process is that it is more environmentally friendly and safer than traditional viscose production methods. It eliminates the use of carbon disulfide, which has negative effect on human health and the environment. With the ability to recycle cotton into novel and sustainable fibers, Infinited Fiber is making significant strides in the development of a more sustainable textile industry.

It should be noted that there are numerous fabrics available that resemble cotton but actually are blends of two or more different fiber types woven together. Polycotton is a popular example of such a blend, combining the softness of cotton with the durability and low maintenance that polyester gives.

To effectively recycle both components, the process must separate them. This can be done chemically by swelling the polyester fibers to eliminate any contaminants, such as dyes and other impurities. This helps to ensure that only clean and high-quality fibers are obtained. Moreover, other polymers in the fabric including polyurethane and cellulose acetate are also washed out. After the washing stage, the pure polycotton is soaked in a solvent until it is saturated. The mixture is then heated to dissolve the polyester fibers, leaving behind cotton. This is followed by filtration for separation. The solvent can then be removed from the polymer and restored for reuse. The recycled polyester fiber can be spun to new yarn. To dissolve the solid cotton fibers, ionic liquid is typically used producing a pulp that is similar to wood pulp. This can then be used as a raw material in conventional cellulosic fiber spinning processes. Some of the examples of fabrics that are presently manufactured using wood pulp include viscose and lyocell. It should be noted that in order to achieve widespread adoption of chemical textile recycling technologies, it is essential to have economies of scale. The feasibility of chemical recycling is therefore dependent on the plant's throughput to produce significant quantities of output per year and to make the cost of manufacturing comparable to current alternatives.

References

Benvus JM (1997) Biomimicry: innovation Inspired by nature. Morrow, New York
Ehrenfeld JR (2004) Can industrial ecology be the science of sustainability? J Ind Ecol 8(1–2):1–3
Hawken P, Lovins AB, Lovins LH (2013) Natural capitalism: the next industrial revolution. Routledge
Hargroves K, Smith MH (2006) Innovation inspired by nature biomimicry. Ecos 129:27–30
Hauschild MZ (2018) Introduction to LCA methodology. Life Cycle Assess Theor Pract 59–66
Lovins AB, Lovins LH, Hawken P (1999) A road map for natural capitalism. Harvard Bus Rev 77:145–161
McDonough W, Braungart M (2010) Cradle to cradle: remaking the way we make things. North Point Press
Osterwalder A, Pigneur Y (2010) Business model generation: a handbook for visionaries, game changers, and challengers, vol 1. Wiley
Pradhan P, Costa L, Rybski D, Lucht W, Kropp JP (2017) A systematic study of sustainable development goal (SDG) interactions. Earth's Future 5(11):1169–1179
UN DESA (2023) The Sustainable development goals report 2023: special edition - July 2023. New York, USA: UN DESA. © UN DESA. https://unstats.un.org/sdgs/report/2023/

Chapter 5
Technology Platforms–Carbon Neutral Technologies

Abstract In this chapter, we explore the crucial intersection of business models and carbon-neutral technologies that offer viable pathways towards achieving carbon neutrality. Many of these technology platforms are already mature or commercially available, making them feasible options for implementation. The six widely recognized climate technology platforms we focus on are electrification; emphasizing the transition from fossil fuel-based power sources to electricity, carbon-free and renewable energy, leveraging hydrogen or ammonia as clean energy carriers, carbon capture technologies and Industry 4.0 Technologies for carbon neutrality including artificial intelligence (AI) and Internet of Things. Throughout this chapter, we provide an in-depth review of each technology platform, exploring their functionalities, benefits, and potential applications. By understanding these technology options, businesses can make informed decisions and integrate them into their strategies for achieving carbon neutrality.

Keywords Technology platforms · Carbon neutral technologies · Electrification · Transportation and city planning · Industrial operations · Hydrogen economy · Carbon capture, storage and utilization · Industry 4.0 technologies for carbon neutrality · Carbon free and renewable energy

5.1 Background

The earlier discussed sustainable business models are imperative in achieving a carbon–neutral society in a global scale. A business model used for driving carbon zero targets must be able to carry the investment cost (return on investment—ROI) and the incurring operating costs. A feasible business model is, however, in most cases only able to take the company so far in corporate carbon neutrality. It is the combination of appropriate and strategically aligned carbon–neutral technologies or technology platforms that are viable options in transitioning to carbon neutrality. These technology platforms or specific technology solutions are mostly mature or

M. Hakovirta, *Carbon Neutrality*, Springer Climate,
https://doi.org/10.1007/978-3-031-45202-4_5

at least commercial. It can be argued for example that the world has already developed affordable low-carbon electricity such as wind and solar, with a very close to comparable price as natural gas or coal. This "green" electricity then connects to the opportunities for electrification, replacing the need of fossil energy resources use for powering many technologies. However, there exist various applications including heavy vehicles, marine time vessels, long-distance trains, concrete, and steel manufacturing that demand major technological innovation to advance carbon zero. These, together with carbon capture and utilization, hydrogen economy and many infrastructure developments are needed and demand major investments not just in their development but also in education and public–private partnerships for successful implementation.

These technologies are critical in solving the grand challenge of carbon neutrality. From a standardization and compliance perspective they also have resemblance to a concept used in the European Economic Community law already 1984, with Directive (84/360/EEC). Since then, the concept of best available technology (BAT) has been used to protect the marine environment in the North-East Atlantic. Later, it has evolved into Integrated Pollution Prevention and Control (IPPC) Directive (96/61/EC). The directive has progressed to the Industrial Emissions Directive (2010/75/EU) governing industrial technologies ranging from the steel manufacturing to the pulp and paper and wood-based panels production. The BAT has therefore been refined into meaning of the most effective and advanced technologies (including design, built, maintenance, operations, and decommissioning) that achieve emission limit values and all permit conditions for reduces emissions and overall environmental impact. The BAT also includes the concept of techno-economic scale applicable to the specific industrial use and appropriate scalability.

This similar approach is used in the USA, and BAT (or comparable term) is used in the Clean Air Act and Clean Water Act. In this respect, the "best available carbon–neutral technologies" or "climate technology" platforms are like BAT technologies that enable the environmental benefits (in this case carbon emissions reduction). The difference is that these new technologies are missing a comprehensive third-party certifications or validation for their impact. Technology suppliers are using their own or consulting services-based Life Cycle Assessments or in some cases industry associations or third-party accreditations in order to give their customers a fact-based proof that they are able to supply the necessary engineering systems that are equipped to fulfill the design criteria for strict carbon zero targets. In some cases, these technologies are not fully mature and demand continuous investments in R&D, demonstration and pilot system investments and public–private collaboration for risk reduction and creating incentives for the development and deployment of such technologies. To all of our benefit, public funding is already focusing on supporting research on these carbon zero technologies, demonstrations, semi-commercial capacity build and start-up company support. In addition, the most impactful technologies and business models are quite well known, and their science and engineering development efforts are on a good track.

Based on various sources in scientific literature, published books, discussions with corporations, start-up companies' investors and funding agencies, the six identified and widely recognized *carbon neutral or climate technology platforms* include *electrification, carbon-free and renewable energy, hydrogen or ammonium platforms, carbon capture technologies, and artificial Intelligence* (AI). We will now focus on reviewing some of these technologies more closely.

5.2 Electrification

5.2.1 Basics

In principle when we talk about electrification, it refers to the transitioning from using technologies that utilize fossil sources as the principal energy source to technologies that use electricity. Evidently, the electricity generating technology portfolio defines the true carbon neutrality of this transitioning (carbon intensity of electricity). For example, in Norway where electricity is mainly produced by hydropower (in 2021, 90% of Norwegian electricity production with 1690 hydropower plants) and windfarms (in 2021, 10% of Norwegian electricity production with 53 Windfarms), the carbon emissions reduction potential can be high. However, in countries or regions where the fossil fuels are the country's top electricity production sources including India with 44% electricity being produced by coal, 24% by oil and 22% by biomass (2020) the positive impact from electrification can be more questionable.

As discussed earlier, in order to achieve the highest positive change, the transitioning to full electrification makes most sense in the industry sectors with high emissions and where electrification is clearly a techno-economically feasible route including *transportation, building sector and certain industrial operations* (Nadel 2019). As always intelligent choices need to be made and the driver for this transitioning is also based on economics and energy security. Some of this transitioning has been swift especially in the personal EV space, stimulated recently by the rising energy crises with record inflation and highest oil and gasoline prices in decades.

The building energy efficiencies and electrification concepts have been developing for years and several frameworks for developing energy-efficient and cost-saving buildings that are also healthy for the occupants have been developed and implemented. A good example of this is the US LEED certification that is a green building certificate by non-profit US Green Building Council (USGBC). The program began 1993 and has since then become the largest green building rating system in the world, including more than 110,000 certifications in about 160 countries (2021). The benefits in LEED are both economic and environmental and create long-term life cycle perspective to the building life and building health. Most impactful target in electrification is without doubt heating and cooling of buildings.

Another high impact area in electrification is the industrial operations that have been reported to have about 35% electrification rate. The *industrial electrification* can be categorized into *industrial fleets, industrial processes, electrification of industrial space,* and *heating of industrial water.* Fleet electrification includes a wide range of transportation systems including *heavy vehicles*, such as various *trucks* and *vans*, *tugs*, and *forklifts*. These are not new developments as the electrical versions of these vehicles are typically less expensive to operate and maintain. However, with the advancement of battery technology and related infrastructure the investment costs have gone down, and the value proposition of cost-savings and environmental benefits have become more appealing.

In terms of industrial processes, *industrial motors* are one of the key areas of electrification advancement. This includes both expanding the electrical motor usage and increasing energy efficiencies. There are also various other areas of high fossil energy use industrial processes that are being redesigned for electrical alternatives including industrial drying. For the industrial space *electrical heating and cooling, increased use of heat pumps* and *smart building management systems* for warehouses, offices, and manufacturing space are some of the most apparent opportunities to implement.

5.2.2 Transportation and City Planning

Electrification of transportation was one of the key topics covered in the 26th United Nations Climate Change conference (COP26), also called Glasgow Climate Talks (2021). In the meeting, 197 countries participated and topics covering for example greenhouse gas emissions, sustainable finance, transportation, cities and urban planning, energy transitioning, climate science, and gender equality. One of the focused topics was electrification of transportation as it is from public opinion perspective considered as one of the key solutions to GHG mitigation. Key achievements from the meeting were a multilateral agreement, in which public and private sector, did pledge to transition to only sell zero emission new cars and vans by 2040. Additional regulatory targets have already been set in the EU and the USA to reach EV share of at minimum 50% by 2030. Also, there are several countries in the so called "leading markets" that have accelerated their timelines for banning internal combustion engine (ICE) vehicle sales 2030 or 2035. These countries have targeted reaching all sales of new cars and vans to be zero emission by 2035. There are also at least 15 countries that have given a pledge to having all sales of new trucks and buses zero emission by 2040.

Electrification of other transportation modes such as aviation and maritime is still far away from realization. The limiting factor is clearly the feasible travel distance due to the energy density difference between jet fuels (12,000 Wh/kg) compared to batteries (lithium 260–270 Wh/kg). There needs to be disruptive battery technologies coming to the market in order to compete with the jet fuel powered engines. Therefore, the aviation industry is mainly focused on renewable fuels solutions such as the Sustainable Aviation Fuel (SAF) (Chang et al. 2020). Rail services, especially in urban areas, are already electrified and enjoy the reputation of sustainable transportation. Maritime is also looking for alternative and more sustainable energy concepts. Recent developments include use of alternatives to maritime diesel power including liquefied natural gas (LNG), hydrogen, or electric energy. The use of these alternative power sources is imperative as it has been estimated that more than 90% of the world trade is using shipping as the means of transportation. There are more than 90,000 ships that contribute to the estimated 3% of the global CO_2 emissions. Moreover CO_2, maritime vessels emit nitrogen oxide (20–30% of global nitrogen oxide emissions) and dust and carcinogenic particles. Just like the aviation electrification, the maritime solutions demand major advancements in the battery energy density. Currently, hybrid solution is to use diesel-electric drives in which diesel engine generated the electricity to drive the otherwise electric engine and propulsion system. Also, LNG-based similar systems and combinations of solar, hydrogen, and fuel cell technologies are used to test new energy approaches.

Cities or urban environments are in a critical role in moving forward current and future electrification concepts. This is because they drive the concentration of population and commerce; they also build and maintain infrastructure that enables electrification of everyday lives. In addition, cities are many times the locations for electric power utilities that are publicly owned entities managing and operating power generation, transmission, and electricity distribution locally and also selling and purchasing electricity from the regulated market. It can be argued that cities are therefore in the core of the electrification as concept testing platforms and players in the energy market. It is worth noting that there are 4 billion people living in urban areas and that the most economic activity globally comes from cities (80% of GDP), they are accountable for almost 80% of the global primary energy consumption and most of the carbon dioxide emissions (75%) (Sustainable Urban Systems Subcommittee 2018).

The concept of *urban electrification* is twofold and includes a practice of providing access to the vast population in urban environments that still are lacking access to it. On the other hand, it is about how to increasingly use electricity as a substitute to fossil fuel-based energy in transportation, buildings, and everyday life. Additionally, it can create an opportunity for developing new cities that have not yet invested in the fossil fuel-based energy and transportation infrastructure. Infrastructure development examples for urban electrification include electrification of transportation and connection of EVs and governmental, commercial, and residential buildings to smart grid, for efficient charging of personal, public, and commercial electric vehicles. This integration also includes concepts of *photovoltaic panels* (PVs) integration to building infrastructure and related *distributed energy resources* (DERs) such as natural gas turbines, *advanced small modular nuclear reactors* (SMRs), wind turbines, biomass-based energy sources, fuel cells, battery storage, and demand response applications. The urban electrification connects also to rural land use as for example wind, solar, or hydropower demand beyond city boundaries, regulatory and continuous building of community relationships for creating clear regional benefits for urban electrification.

5.2.3 Industrial Operations

Industrial electrification provides a major opportunity to CO_2 emissions mitigation. Based on a recent study, 78% of the industrial energy demand can be electrified. Most interestingly, this can be achieved by using established technologies that are available. Looking forward into the future, pushing the envelope can create 99% electrification with technologies that are still being developed and emerging. This level of electrification will also create reduction in final energy consumption of 20%, corresponding to 9% overall CO_2 emissions reduction using the current energy mix for electricity production (Madeddu 2020). The referenced study also finds that 78% industrial electrification levels can create even 50% CO_2 emissions reduction from industry if the decarbonization of power supply sources reaches the level of 12 gCO_2/kWh_{el} (current electricity mix is about 300 gCO_2/kWh_{el}). This level is reachable currently' by nuclear, hydro, and wind (solar PV is just below 50 g CO_2 e/kWh_{el}). As a comparison, highly efficient (50–60% efficiency) combined cycle gas turbine (CCGT) plants are at the level of 400–450 gCO_2/kWh_{el}, and therefore, reaching the desired 12 g level for overall electricity production is taking a long time and estimated to take place in 2050.

The electrification opportunities in industry can be divided into heating and cooling, mechanical power, and lighting. In this context, the lighting can obviously be considered as fully electrified. Cooling in the commercial sector in the USA (commercial and institutional buildings) uses approximately 153 billion kWh (2021). It also represents 4% of total US electricity consumption and some 12% of the total commercial sector consumption. Unfortunately, specific data for industrial cooling does not exist for example in the US EIA databases. Obviously, the use of cooling is dependent on the industry sector and the processing conditions needed. For example, for food industry and related transportation cooling is a high-energy consuming activity. As an example, the total global food production is about 4 billion metric tonnes, and an estimated 1/3 requires refrigerated processing. According to the Global Cold Chain Report of 2020, there are approximately 719 million cubic meters of refrigerated food storage warehouses available worldwide.

Another example is Information and Communications Technology (ICT) industry. Its electricity demand is more than 2000 TWh annually and accounts for about 10% of the total global electricity demand (global consumption was 23,900 terawatt-hours). All data centers that are cooled consume 200 TWh annually with an approximately 40% used for cooling by conventional air-conditioning systems or use of cooling towers. Currently, for example, Google has used new approaches to reduce the power consumption of the server cooling including using AI and has managed to drop the electricity use by 30–40% (Bronner et al. 2021; Xu et al. 2019). Both residential and commercial cooling needs are highly dependent on the weather conditions and climate, and therefore with increased global temperatures, cooling is becoming a challenging issue to tackle.

The energy associated with heating relates to space, steam production, and thermal energy that is further categorized by different temperature ranges. For the most part industry sectors that mainly use lower than 400 °C temperatures can be considered as "low hanging fruits" from electrification perspective. Such industries include food, textiles, wood, and paper industry. Mature and existing technologies can fulfill most if not all the necessary electrification needs. Typical examples are compression heat pumps, mechanical vapor recompression systems (MVR), electric boilers, infrared, microwave, and radio frequency heaters. These technologies are immediately available and account for 35% of Useful Energy (UE) for industrial electrification needs. Here, UE is defined as the energy available after the efficiency and energy losses are accounted for from the energy input into the used technological system.

Other industries such as ceramic, steel, and chemicals industry are more challenging sectors and chemicals industry has the highest potential due to its nature and need for cooling and steam generation for cracking and reforming. These more challenging electrification application are above 400 °C temperatures and utilize electric furnaces, resistance heating, induction, and arc furnaces. The list includes electric kilns that can be used for calcination replacing gas operated rotary calciners. Although many of these more advanced technologies are quite mature, these applications demand some advancements in the process side. To reach the full potential of industrial electrification, there are various technological uncertainties.

The low maturity and high-risk technologies have their impact in chemicals, cement, and steel industries and can create a further electrification potential in these industrial sectors. Examples include electric steam crackers and reformers. The actual estimated full energy savings potential being 60% (Madeddu 2020). The rest of the potential is connected to the fossil fuel use for non-ferrous metals smelting, electric arc furnace use, and chemical feedstock needs. Especially from feedstock perspective, a full fossil independence is thus not feasible. If we would not consider the feedstock part, then the actual electrification potential is 99% (Madeddu 2020) (Table 5.1; Fig. 5.1).

Looking more closely at the use of electricity in industrial processes, it includes mainly refining and grinding using mechanical equipment (e.g., mechanical pulp in pulp and paper industry) and the also equipment such as pumps and fans, compressed air systems and various other motor operations. Although electrification in these processes and systems is mature, more work is in progress in continuous improvement of efficiencies and the advancement of systems design for the achievement of total system efficiencies and integration to green power sources in a seamless way.

5.3 Hydrogen Economy

5.3.1 Basics

The Herculean task of transitioning to global net zero energy system is imperative for reaching the goal of keeping our planet's warming at well below the 2 °C compared to pre-industrial levels. This process will take decades and will demand rethinking

Table 5.1 Electrification and industrial heating and cooling

Technology	Technological maturity	Applications
Heat pumps and cooling (<100 Celsius-100 Celsius)	Mature	Heating and cooling air, hot water and steam
Mechanical Vapor Recompression (MVR) (>100 Celcsius - 400 Celsius)	Mature	Energy recovery from process heat
Electric boilers (<100 Celsius - 1000 Celcius)	Mature	Heating, thermal oil, steam
Infrared heaters (<100 Celsius - >1000 Celsius)	Mature	Drying, suring, food processing
RF and microwave (<100 Celsius - >1000 Celsius)	Mature except for materials firing and sintering	Drying, food processing, development phase for materials firing and sintering
Furnaces (induction or resistance) (<100 Celsius - >1000 Celsius)	Mature	Metals heating, melting, smelting, sintering, glass, melting, calcination
Electric arc furnaces (<100 Celsius - >1000 Celsius)	Mature	Refining and metals melting

of business concepts, earning models and readjustment of global investments. The transitioning will redefine how energy is produced, distributed, and stored. One of the key technology platforms in moving this transformation process forward is the use of hydrogen in a large–global–scale. The demand for hydrogen has tripled since its first use in 1970s. Currently, about 94 Mt of hydrogen is manufactured globally. Its main uses are ammonia, polymers, and resins (65%), refining—including hydrocracking and hydrotreating (25%), and iron and steel including general industry use (10%) (Global Hydrogen Review 2022).

At the moment, China leads as the hydrogen producer, with approximately 25 million tons per year. This counts for almost one fourth of the global hydrogen production. However, the Global Hydrogen Review of 2022 states that a significant portion of this volume is derived from fossil fuels, with approximately 60% coming from coal and 25% from natural gas that are used as feedstocks in refineries or chemical facilities.

Hydrogen platform is one of the necessary technologies that makes the carbon–neutral future feasible and the ambitious GHG emission goals achievable. It has a high readiness level to decarbonize several industrial sectors including transportation,

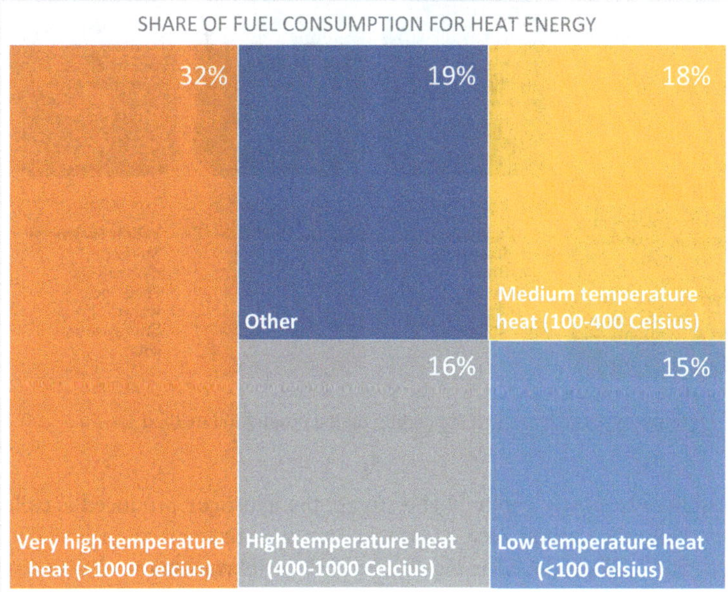

Fig. 5.1 Share of fuel consumption for different industrial heating applications (Madeddu 2020)

heavy industry, and building industry. It also provides the necessary flexibility that all green energy solutions need. For these reasons, demand for hydrogen is rapidly increasing, however, hydrogen is still mainly fossil-based. As low-carbon hydrogen is becoming available, it will be a vital tool for decarbonization for various sectors. Electrical power will also be in an important role as it can be used to produce hydrogen using electrolyzers. The transportation of hydrogen from production site can be efficient to the point of use as it can be done long distance by using an existing pipeline infrastructure and by road in super insulated cryogenic tanker trucks or gaseous tube trailers. This infrastructure also provides the necessary flexibility in green energy supply–demand fluctuations and imbalances. Hydrogen can be stored in old salt mines and depleted gas fields or other appropriate cave structures. Additionally, it can provide a low barrier of adaptations with similarity to combustion engine use or natural gas use in certain applications.

Hydrogen can be categorized to eight types depending on the production method used (gray, black, brown, blue, green, turquoise, pink, and yellow hydrogen). *Gray hydrogen* is formed from natural gas or methane and thus fossil-based. It is the most used production method currently (roughly 95%), and nearly 6% of global natural gas is used for hydrogen production. The production process is called steam reforming. It is based on production of syngas (hydrogen and carbon monoxide) using reaction of hydrocarbons with steam in the presence of a catalyst. The same process using bituminous coal or lignite coal as the hydrocarbon source is called *black or brown hydrogen*, respectively. If the CO_2 that is produced and released to

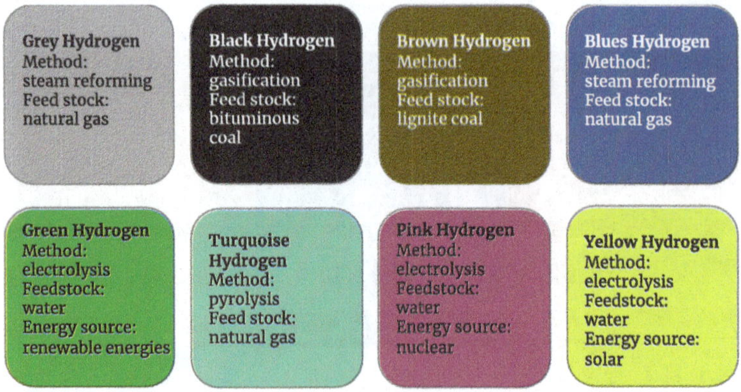

Fig. 5.2 Different types of hydrogen depending on the production method

the atmosphere is mostly captured and stored, the hydrogen produced is called *blue hydrogen*. Now the carbon zero alternative that we are focused on is called *green hydrogen*, and it is produced using the process of thermochemical water splitting in which electrical current (also called electrolysis of water) or high temperature heat (500°–2000 °C) produces hydrogen by decomposing water into hydrogen and oxygen. Green energy sources for these production processes can include solar, wind, hydropower, and nuclear power. Using nuclear energy as a source is also called *pink hydrogen* and electrolyzes with solar power solely is called *yellow hydrogen*. The gray hydrogen clearly cannot be considered as low-carbon alternative as it has even a slightly worse carbon footprint than methane. Blue hydrogen having carbon controls including sufficient CO_2 storage available can be considered as an option to low-carbon fuel. There are only a few production sites globally that are purposed to do the CCS process in this context (Collodi et al. 2017). Lastly, turquoise hydrogen is produced by methane pyrolysis that can directly split methane (from natural gas). The end products are hydrogen and solid carbon. The produced carbon has other uses and is also known as *carbon black*. The various applications of this material include coatings, fillers in plastics, filler in production of tires, and batteries. The carbon black has a global $20 billion market and has prices based on materials characteristics as high as $1400 per metric ton (Carbon black market 2021) (Fig. 5.2).

5.3.2 Heavy Industry Use of Hydrogen

The energy density of hydrogen is high at 33.33 kWh/kg, while petrol and diesel have 12 kWh/kg energy density. However, volumetric energy density of hydrogen is easily overcome by liquid fuels. Since there are challenges in the supply–demand of green hydrogen the limited green hydrogen supply should be used for the highest GHG reduction applications. Some of the industrial processes are the most GHG emitting sources including steel production, chemical industry, and cement production. Also, the heavy industry is energy intensive and uses a considerable amount of energy in various forms. The need for process heat is one of the major reasons for this and an opportunity for hydrogen usage in various applications. Heat is used in multiple process steps including drying of materials, water heating, melting, and sintering of metals and so on. The heat used is coming mainly from fossil fuels utilization including coal, oil, and natural gas, contributing to over 40% of the industrial GHG emissions (IEA 2019).

Depending on the process, GHG emissions can be reduced by electrification and using more green electricity, renewable fuels, or zero carbon fuels like hydrogen. And in some cases, the use of coal or natural gas can be replaced by using hydrogen, especially processes that need high grade heat (higher than 650 °C) or medium grade heat (200–650 °C). This approach has been estimated to reduce total CO_2-eq GHG emissions by about 85 Mt (Sintef 2019). Steel industry can use hydrogen in two ways. One is to use hydrogen in the direct reduction (DR) process in combination with electric arc furnace (EAF). This is also called DR-EAF route. In this process, hydrogen is passed through the shaft furnace, and it strips O_2 from the iron ore (Fe_2O_3), resulting in sponge iron, which is then added to the EAF process, and many times mixed with scrap steel as well. The EAF process electrodes will use current to melt the sponge iron to produce steel. Still in the steel manufacturing process carbon is needed, but it can be added by using biomethane, pulverized coal, or other biogenic carbon sources. Also, the mixed scrap metal can reduce the need of the added carbon. The GHG emission reduction compared to the traditional process can be as high as 95% (Hydrogen roadmap Europe 2019). The DR process can also be used in combination with other steel manufacturing processes such as blast furnace or basic oxygen furnace (BOF). This creates necessary heat and strips the oxygen from the iron ore. Using fuel mix approach will leave some of the emissions, and thus, emission reduction will be limited to about 20% (Yilmaz et al. 2017).

Hydrogen is fundamental for chemical industry. It is used mainly for the manufacturing of ammonia and thus fertilizers. Also, it is critical for manufacturing methanol, which is used for many polymers. More than 80% of the hydrogen that is used in chemical industry is for the production of ammonia, 10% for the production of methanol and 2% for both polyurethane and nylon. The key to decarbonize the ammonia production is related to the type of hydrogen used. With gray, brown, and black hydrogen the process contributes to GHG emissions, and it has been estimated that using green hydrogen can reduce emissions by 30 Mt of CO_2-eq. The challenge here is therefore the available supply of green hydrogen and it has been estimated that to replace the gray, brown, and black hydrogen with the green alternative would require globally 173 TWh of green electrical power if electrolysis is used (Agora Energiewende 2021). This points out to the importance of the other alternatives such as pink and yellow hydrogen production. Interestingly, hydrogen is also used for *hydrodesulphurization* (*HDS*) which is one of the catalytic chemical processes for the removal of sulfur from natural gas and various other more refined petroleum products including gasoline, jet fuel, and diesel. Transitioning to green hydrogen in this respect would also improve GHG emissions in the oil and gas industry.

5.3.3 Hydrogen-Based Transportation Technologies

The EVs have recently become a popular green alternative solution to personnel and light vehicle transportation. However, the heavy vehicles and trucks are still looking for a more sustainable solution. The new GHG emissions reduction targets have strict emission rules to heavier on-highway trucks and for example in the USA the target is by 2027 to reduce these CO_2 emissions by 46% from 2010 levels. In addition, a coalition of fifteen states is requiring having 30% of the new trucks on the road to be carbon zero. Similarly in EU, the CO_2 emissions of new heavy vehicles and trucks are regulated to have 30% cut by 2030 with 2019 as the baseline (Regulation (EU) 2019/1242). China has also set a goal of using over 1 million hydrogen-powered commercial vehicles by 2030. These regulations do not include the off-highway heavy vehicles, however, as mining companies' operations scope 1–2 emissions total about 30% and are coming mainly from mining heavy vehicles, major mining companies including Rio Tinto, BHP, and Anglo American are now working on their off-highway heavy vehicles transitioning to carbon zero.

The hydrogen-based carbon zero technologies for the heavy vehicle and truck use include *Hydrogen Fuel Cell Electric Vehicles* (*FCEVs*) and *Hydrogen Internal Combustion Vehicles* (*H2-ICEs or HICEV*). The FCEVs are as any battery operated EVs and use electric power for motion carried out by electric motor. The difference is that in the FCEVs the power is not coming from a battery bank; it is generated by a fuel cell that uses hydrogen as the power source. Thus, the FCEVs are considered to have lower environmental impact than EVs. These electric vehicles also include battery system for recapturing and releasing breaking energy for low speed and acceleration needs. Structurally, fuel cells main components are membrane, catalyst layers (anode and cathode) and diffusion media. The membrane is typically, polymer or ceramic membrane material and is less than 20 microns (20×10^{-6} m) in thickness. It conducts or allows only positively charged hydrogen ions to pass through to cathode. Catalyst layers are on both sides of the membrane and are typically only a few nanometers thick platinum particle layers on a carbon support and also bonded with ion-conducting polymer. The anode side catalyst causes the hydrogen gas atoms to split into positive hydrogen ions (protons) and electrons. The electrons then travel by means of external circuit from anode to cathode and thus form the electric current. In the cathode, the catalyst allows oxygen reaction by combining electrons and positively charged hydrogen ions with oxygen (from air). This then creates the by-products of pure water and heat. These are the operating principles, but the actual system is much more complicated and includes for example fuel cell stacks, fuel processors, power conditioners, air compressors, and humidifiers. One single fuel cell produces only less than 1 V of difference of potential, and therefore, there needs to be hundreds of fuel cell stacks in the power system structure to be able to power a vehicle.

The advantages of hydrogen fuel cells include the fuel efficiency, short fueling time (a few minutes), long range and carbon zero approach to heavy vehicles. Disadvantages include high capital cost of the vehicles, lack of charging and service infrastructure, safety concerns, and green hydrogen availability. One major drawback, as mentioned earlier, is also that hydrogen has a low volumetric energy density and therefore demands high-pressure tanks in low temperatures in order to keep hydrogen in liquid state. When liquefied its volume will be reduced by a ratio of 800–1, the drawback is that the temperature needs to be kept at minus 253 °C. In the future, hydrogen metal or non-metal hydrides are forecasted to replace heavy hydrogen tanks.

Hydrogen Internal Combustion Vehicles are in principle modified versions of the gasoline-powered internal combustion engine. It is a carbon zero vehicle since the fuel has no carbon in it, and therefore, no carbon monoxide (CO), CO_2, or hydrocarbons (HC) are created. One benefit of the hydrogen engines is that they can operate with lower grade hydrogen without the need for purification (unlike with fuel cells). This reduces the cost and flexibility of using various types of hydrogen as discussed earlier. Although the technology is established, there are new technical concepts that can be used for high power applications. These can alter from direct injection to port injection and from compression ignition to spark ignition. On the whole, the main benefit of combustion engines using hydrogen is that they can produce torque to power wheels or propeller directly with no need for conversion to mechanical power such as in the case of fuel cell produced electricity. They also perform better at higher loads with less efficiency losses due to auxiliary components like in fuel cell systems. Also, when more power is needed the fuel cell needs to be larger and thus more complex. In case of combustion engine, scaling up creates more efficiencies.

Maritime transportation industry operates ports and waterways and is responsible for global shipping of commodities such oil and gas, agricultural products, automobiles, and home appliances. It is estimated that about 80% of global trade (by volume) is transported by ships at sea. Marine transportation is also important in transporting passengers especially in regions with islands and cross-water traffic. All this is done using large containerships, cruise ships, and ferries, and most are powered by heavy diesel producing globally very large quantities of air pollutants that include nitrogen oxides (NOx) and sulfur oxides (SOx). It is estimated that about 15% of NOx emissions and 13% of SOx comes from marine transportation (UNCTAD 2017). In terms of CO_2 emissions, maritime transportation contributes to approximately 3% of global greenhouse gas (GHG) emissions (measured in CO_2-equivalents). This equates to an average of around 1 billion tonnes of CO_2-equivalents annually. The International Maritime Organization (IMO) has joined the efforts in reaching carbon neutrality and has committed to 40% GHG reduction from 2008 level by 2030 and 50% by 2050 (IMO GHG Strategy 2018). These targets are challenging and to reach them demands a variety of design, operational, and economic solutions. In terms of potential fuel or energy usage and propulsion technology in maritime transportation, there will be changes towards using for example biofuels, ammonia, electric power, hydrogen, fuel cells, methanol, and wind.

Hydrogen can be used in *maritime* similarly as with vehicles, by H2-ICEs or HICEV technology. The drawback here is that even though hydrogen is zero carbon fuel, when it is burns in air, which is composed of mainly nitrogen (78% nitrogen and 21% of oxygen), it does produce nitrogen oxides that are also targets for emission reduction. Post-combustion treatment can however be a solution to this drawback.

The alternative is the hydrogen fuel cell technology that essentially is emission free. The space needed for storing hydrogen is obviously large and continues to be a concern in the industry. However, latest analysis by for example International Council on Clean Transportation (ICCT) shows that most cargo ships across the Pacific Ocean could be equipped with hydrogen power and storage without any need for additional cargo space for fuel. There are other options including hydrogen fuel mixing or hydrogen fuel enhancement by *adding hydrogen to hydrocarbon fuels* in an internal combustion engine and thereby improving the power output and reducing the emissions. In case of *adding hydrogen to diesel*, the environmental benefits include more complete combustion and even 15–20% improvement in fuel economy. One solution that is more specific to NOx and SOx emission reduction is to use liquefied natural gas (LNG) only. In principle, it can reduce CO_2 emissions by about 20% compared to maritime diesel (it has higher hydrogen to carbon ratio). In addition, it reduces up to 80% of the NOx emissions and can also eliminate SOx and particulate matter (PM) emissions. Unfortunately, LNG using combustion engines does not have 100% efficiency (close to 98%), and therefore, some of the methane is released as unburned gas contributing to the GHG emissions. If we account for supply chain losses of the LNG, this combined effect has been estimated to eliminate the positive GHG reduction effect in using LNG.

More advanced solutions for maritime power include using *turbine for electricity generation powered by steam that is generated by burning hydrogen and pure oxygen.* The same technology would work similarly in trains. There is also an ammonia alternative. That is *hydrogen can be used to make ammonia* (NH_3). It is more versatile and can be *combusted or used in fuel cell applications.* The main benefit is that *ammonia does not demand as low temperatures for storage as hydrogen and can be stored in a regular pressurized tank.* It is denser, takes less volume, and *can also be converted to hydrogen using catalyst and membrane technology in the maritime vessel itself*. It can then feed to fuel cells or other power generating technologies we have already discussed (Fig. 5.3).

The issue with *ammonia is* the fact that it is *toxic and can harm both humans and marine life.* Also, there is energy efficiency losses and monetary cost involved in the transforming the hydrogen to ammonia and vice versa. Also, *Liquid Organic Hydrogen Carriers* (*LOHCs*) can be used. This also allows flexibility for the storage and transportation of renewable energy. LOHCs are a type of energy storage medium

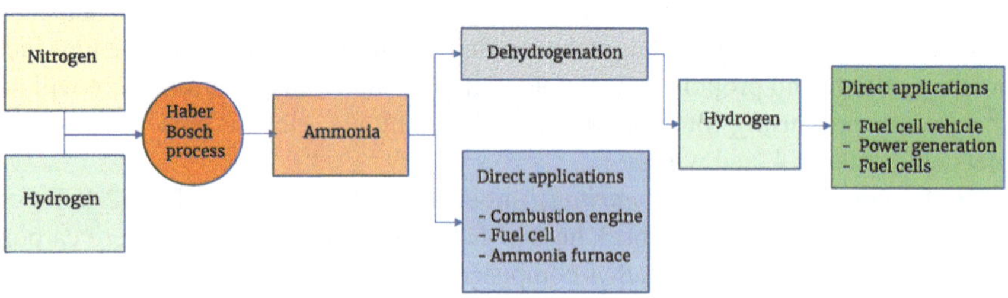

Fig. 5.3 Ammonia as a means of storing, transporting, and using hydrogen

that can be used to store and transport hydrogen. LOHCs are organic compounds such as toluene or naphthalene that can absorb and release hydrogen through reversible hydrogenation and dehydrogenation reactions. This makes LOHCs function as "liquid hydrogen batteries" that can store hydrogen in a stable and safe form at ambient temperature and pressure and release the hydrogen on demand when needed. LOHCs have the potential to play a key role in the transition to a hydrogen-based energy economy, as they offer a convenient and efficient way to store and transport hydrogen, which is otherwise difficult and expensive to store and transport in its gaseous form.

Another interesting alternative would be to use methanol or methane as fuels produced by hydrogenating pure CO_2 with H_2 with nickel, alumina-ruthenium, or Cu/ZnO-based catalysts. Both alternatives can function as fuel for combustion and fuel cell-based powertrains. With strict 2030 and 2050 targets in both heavy commercial vehicle and maritime industry, there is a major technology transitioning that will regardless of the technology selection demand major capital equipment investments. Many of the benefits of these hydrogen-powered technologies call for thorough understanding of the new technological platforms, and operating parameters to be able to choose the hydrogen technology that fits the specific transportation application.

Aviation is another potential hydrogen technology users. With an estimated 900 million tons of CO_2-eq. emissions and a continuous increase in demand for air travel, the emissions will more likely double by 2050 even if we factor in continuous efficiency improvements (Hydrogen-powered aviation 2020). The fact that the aviation industry (Air Transport Action Group—ATAG) has committed to 50% GHG emissions reduction by 2050 (2005 level as the reference) has created a tremendous momentum to find solutions as well as concerns of the validity of the targets in the chosen timeframe. Similarly, EU has even stricter targets in its Green Deal with an ambitious goal for all industrial sectors, including aviation, reaching carbon neutrality by 2050. In addition to CO_2 emissions, aircrafts also emit NOx, water vapor, and soot into the atmosphere. The associated affects include increase in ozone emissions from NOx emissions which has been estimated to have even comparable to the global warming effect of CO_2 (Sausen and Schumann 2000). These additional other climate change-related effects increase the impact of aviation to an estimated 3–7% of total CO_2-eq emissions (2–4 gigatons of CO_2-eq).

The history and reputation of hydrogen in aviation is somewhat gloomy. On May 6, 1937, hydrogen filled *Hindenburg disaster* occurred in New Jersey, USA. The German passenger airship also called Zeppelin (invented by Count Ferdinand von Zeppelin), while docking, came down in flames and explosions. This catastrophic event marked an end to the hydrogen-powered passenger-transporting era for decades. Even with major developments in hydrogen technology, main challenges for the future of hydrogen in air travel remain to be the infrastructure. Regardless of the power delivery system, the transportation of hydrogen to the airport in a safe and functional way including fueling remains somewhat of a challenge. Hydrogen propulsion technologies for aviation currently include two main approaches: *hybrid and battery electric aircrafts and direct hydrogen-powered aircraft*. The main challenge in the electric powered technologies is that the battery technology is still giving only up to 0.5 kW-hours per kg in terms of energy density. Therefore, weight and the lithium battery life cycle length will be a major barrier for long range aviation application. The hybrid approach would include *hydrogen fuel cell-based added power* including *power generated by established propulsion technology such as electric motor and propeller or fan.* If the battery technology was solved, the infrastructure changes for aviation industry charging systems would also mean major investment issue moving forward. A true hydrogen aircraft uses a hydrogen burning turbine engine to create thrust, or it can also use hydrogen fuel cell technology. As with other transportation applications, the issue of high storage volume using hydrogen demands larger fuel tanks that also need to be kept at high pressure and low temperature (20 °K). This is a major hurdle for long-distance flights and the overall techno-economic feasibility of the aircraft. Some NOx emissions will occur in burning hydrogen but with for example Selective Catalytic Reduction (SCR) it can mainly be avoided (Borchers et al. 2021). As with combustion engine applications with turbine engines ammonia can be used as fuel to reduce the cost and risk associated with fuel storage. The ammonia fuel would be stored in the wings just like kerosene and pumped through a heat exchanger to heat it. Then, a catalyst can be used in a reactor to separate hydrogen and nitrogen, and the mix is burned in the combustion chamber of the turbine system. This approach would better allow the also of existing aircraft and engine designs, potentially reducing any switching costs. Regardless of the approach, fuel price is one of the key items in an airline's cost structure. The economy of scale needs to be reached in hydrogen economy to break this significant barrier for utilization of hydrogen fuels.

5.4 Carbon Capture, Storage, and Utilization

5.4.1 Basics

Energy use related to operations in industry, transportation, and buildings are estimated to account for about 57.9% of all global CO_2 emissions (Our World in Data 2021). The source of the used energy (power generation) and energy intensive and most CO_2 emitting manufacturing operations (e.g., steel and cement manufacturing) are challenging to transition to carbon neutrality. Some of the possible technological solutions we have already discussed, however, an approach to capture CO_2 prior to entering the atmosphere demands specific attention. In this process, CO_2 is also compressed, transported, and finally stored for longer periods of time or directly used in various industrial applications. The idea of capturing carbon from its source is simple and powerful. On the other hand, it can also be seen as the escape out of the crises, and a reason not to work hard on the highly complex engineering developments, making high capital investments, and taking risks in new business models for long-term benefit of changing the course of climate change.

The carbon capture technology is not a new invention and was already proposed as early as in 1938. The first larger scale system was implemented in Texas in 1972, followed by a first ever offshore Carbon Capture and Storage (CCS) system installation in a Statoil owned and operated natural gas field in Sleipner, Norway. The motive was clearly economical, and the CCS process allowed the company to not to pay the Norwegian CO_2 tax by obtaining carbon credits for injecting the CO_2 to the ground (Utsria formation). This area is a massive sandstone formation (250 m thick) and can potentially store 600 billion tons of CO_2. Currently, worldwide, there are 31 pilot and demonstration CCS plants that are operational and 29 commercial CCS facilities (Global CCS Institute 2023). There are five new pilot and demonstration plants under construction and eight new commercial systems currently being built. The Global CCS Institute collects this data and is by its purpose an international think tank purposed to accelerate the deployment of CCS. They define pilot CCS plants as operational for technologically capturing, testing, and piloting CO_2 capture with possible storage capability. Also, these systems are not necessarily looking at the storage concepts nor are they expected to create commercial validity. Furthermore, the actually commercial CCS facilities are on the other hand specified as having operations to capture CO_2 and to transport it to a permanent storage location. They are also part of the specific company's business model and general operations for long term. These locations are integrated to different industrial application needs including for example: natural gas processing, power generation, chemicals processing, iron and steel production, oil refining, and fertilizer production. Globally, each region has different CCS readiness levels, including legal and regulatory maturity, and needs to deploy CCS for the CO_2 emissions mitigation. This can be affected by the overall regional fossil fuel use or existing industrial activities.

There are basically three generations of CCS technology concepts including *pre-combustion, post-combustion* (PCC), *and oxyfuel capture process.* The *first-generation approach* has been scrubbing *CO_2 using* nitrogen-containing organic compounds; *amines*, and the *second- and third- generation technologies* use *carbonate looping* or *chemical looping combustion.* These different methods separate CO_2 from gas mixtures using chemical or physical absorption, membrane separation and chemical looping.

5.4.2 *Precombustion CCS Technology*

In precombustion CCS a carbon containing fuel (e.g., coal, natural gas) is refined prior to combustion. This is performed in order to create both CO_2 and H_2 rich gas. The next step is to separate the CO_2 and use the remaining H_2 as the actual fuel. In order to do this, the carbon containing fuel is gasified in high heat and pressure and also with steam. During the combustion process, oxygen is supplied only enough to create temperature that can decompose the fuel and thus create the synthesis gas (syngas). The components of syngas are carbon monoxide (CO) and H_2 and small quantities of other gases, such as methane. A water–gas shift (WGS) reactor is then

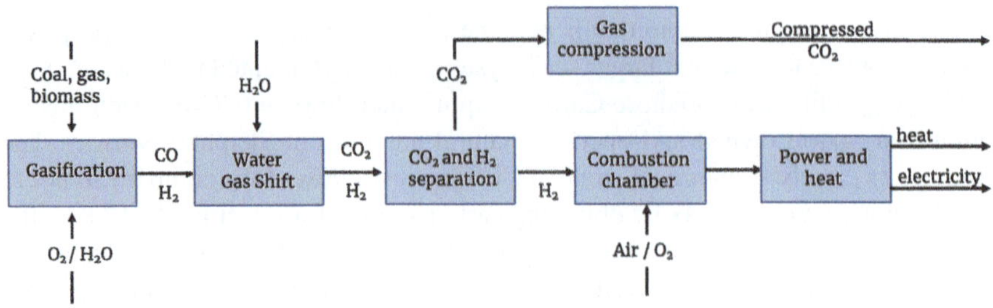

Fig. 5.4 Precombustion CO_2 capture method block diagram for heat and electricity generation

used to convert CO into CO_2 by using steam present in the syngas, thus creating higher CO_2 concentrations in the mixture (typically at the level of 15–50%).

At these levels, CO_2 can be then separated, compressed, transported, and sequestered or used in commercial applications. The benefit of this process is that the CO_2 gas stream is with higher concentration and pressure than with the post-combustion technology, thus allowing a more efficient separation of CO_2 prior to H_2 combustion. It is estimated that even as high as 92% of the manufacturing plant's CO_2 emissions can be captured (EPA 2022).

The flaw is that the capital cost of the plant with gasification process capability is high. The drawback here lies in the high capital cost associated with a plant equipped with gasification process capability. Moreover, the operating costs for carbon capture technology, whether using physical or chemical adsorption processes, are usually estimated to range between \$50 per ton of CO_2 and \$300 per ton of CO_2 (Kazemifar 2022). Further development of the separation technologies using for example advanced solvents, sorbents, and membranes are done in order to push down the cost to a targeted \$30 per tonne CO_2 level (Kazemifar 2022; Ganguly et al. 2022). Some of the drawbacks of this technology include the drop in overall plan energy efficiency by about 20–30%. In addition, the operating costs are increased by an estimated 15–20%, and the capital costs of the equipment and installation are highly capital intensive (Kazemifar 2022). As an example, the Boundary Dam 824 MW coal-fired plant, located in Estevan in Saskatchewan, Canada, invested in CCS precombustion technology. The CCS part for this plant covers 110 MW of base-load electricity with the target of reducing CO_2 emissions in an annual basis by one million tons. An estimate of the retrofit part of the project was about \$354 million and the cost of the CCS \$1.2–1.4 billion (Carbon Capture and Sequestration 2022) (Fig. 5.4).

5.4.3 Post-combustion

The post-combustion method is distinct from the precombustion from the fact that it captures CO_2 from the gas streams that are produced by the combustion of fossil fuels

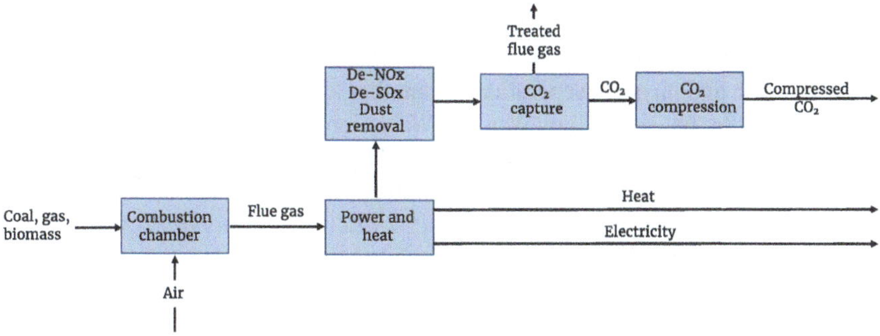

Fig. 5.5 Post-combustion CO_2 capture method block diagram for heat and electricity generation

or other carbon containing feedstock. In this technology the purpose is to remove dilute CO_2 (5–15% concentration) from low pressure flue gas streams. The actual installation of these systems is in the section that follows the flu gases that have been purified of sulfur, nitrogen, and dust. Although, this technology is the mostly considered approach to be installed in conventional power plants, the capturing of the CO_2 can be challenging and the post-combustion concept can be divided into different methods including using absorption solvents, adsorption and physical separation, membrane separation, *chemical looping combustion* (CLC), *calcium looping process* (CLP), *and cryogenics* (Fig. 5.5).

The absorption solvents method is based on chemical reaction between the solvent and CO_2. Alkanolamines are typically used as the solvent and more specifically aqueous solutions of *monoethanolamine* (MEA), *diethanolamine* (DEA), *and methyl diethanolamine* (MDEA) (Porcheron et al. 2011). In addition to these traditional solvents, also new solvents and solvent blends are being developed. Some of these examples include *ionic liquids* (ILs), which can both physically and chemically absorb CO_2. Also, water-free solvents, that reduce energy use in the stripper and biphasic solvents, that can increase CO_2 capture efficiencies are being developed. One of the latest advancements is the *Deep Eutectic solvents* (*DESs*) that are similar in properties as ILs, however, they are environmentally more friendly and provide a lower cost option to ILs.

In the post-combustion method, the flu gas is passed through the absorber column in which the reaction occurs, and the CO_2 is captured, followed by the regeneration of the CO_2 in a stripper section using high temperatures. The CO_2-free solution is then used again in the absorber, and the stripped high purity CO_2 stream is compressed, transported, stored, or utilized in various applications. This process is quite established and can capture 85–95% of the CO_2 in flu gas with very high purity (Asif 2018). The technology unfortunately uses a lot of heat energy in the stripper or the regeneration process. A physical absorption method is also used by applying chemically inert solvents. This approach seems to work best with CO_2 capture in coal gasification processes operating in lower temperatures and high pressures. There are several commercial solvents in the market for various CO_2 and other gas removal purposes that can be used (e.g., Purisol™, Sulfinol™, Selexol™).

Adsorption and physical separation are another technique to capture and separate CO_2. It relies on physisorption (Wan der Waals forces), or chemisorption (ionic or covalent bonding). In the case of CCS, the CO_2 is removed from the gas mixture and adsorbed physically onto various materials that are used including zeolites, *metal organic frameworks* (MOFs), *metallic oxides* (MO), *mesoporous silica, porous organic polymers* (POP), activated carbon, and alumina.

Typically, the process includes a column that is packed by these spherical adsorbent materials and gas stream with CO_2 is passed through it. The CO_2 then flows through and is in contact with the adsorbent material and adheres on the surface by the adsorption forces. Four methods commonly employed for this process include pressure swing adsorption (PSA), temperature swing adsorption (TSA), pressure–temperature swing adsorption (PTSA), and vacuum swing adsorption (VSA). Under high pressure, gases are prone to be adsorbed onto solid surfaces and this phenomenon is the principle of PSA. With increasing pressure, the quantity of the adsorbed gas is increased. Consequently, as the pressure is dropped, the adsorbed gas is released, or desorbed. Two pressurized adsorbent vessels are used to create an almost continuous production of the target gas from the gas mixture. Also, pressure equalization, during depressurization can be used to also help pressurize the second vessel, thus creating energy savings in industrial operations. TSA separation process is using thermodynamic characteristics of the selected adsorbent material.

At low temperatures, certain materials adsorb selectively certain gas than at higher temperatures. In the CCS process using TSA, flue gases are introduced to a fixed bed of CO_2 adsorbent material (e.g., zeolite) that selectively adsorbs CO_2 until the saturation level is reached. After this, the flue gases are redirected, and fixed bed is heated up to the desorption temperature in which CO_2 is released and the system is purged from the CO_2, followed up by a cooling of the system for the next adsorption cycle. The PTSA is a combination of both PSA and TSA with a purpose of gaining higher efficiencies that is still under investigation. The VSA on the other hand is based on using an elevated (higher than atmospheric) pressure, while desorption step is performed at vacuum pressure.

Membrane gas separation (MGS) is a technology used for the separation of specific gas components using the difference in their permeation rates through a thin membrane barrier material. There are several membrane types including organic (polymers) and inorganic membranes (e.g., ceramic, metals, carbon, silica, zeolite, alumina, titania, and zirconia) and hybrid membranes. All these materials have pros and cons including glassy polymeric membranes having high selectivity but low permeability and aging issues. On the other hand, rubbery membranes have low selectivity but high permeability. In general, the use of polymeric materials in membranes enables processing them into large variety of shapes and forms (including thin films) and also their light weight. Inorganic membranes are much more restricted in their form but are attractive from both selectivity and permeability perspective. They are also more resilient to harsh operating environments with high temperatures and pressures. The challenge in them is their high cost that has been a barrier to commercial applications. The hybrid membranes offer benefits of both polymeric and inorganic membranes. They can be shaped like polymer membranes and are more economical, and they can offer the same high selectivity of inorganic membrane materials.

There are several commercial products for CO_2 separation including Polaris™, PolyActive™, and PermSelect®. One of the latest developments is the metal oxide frameworks (MOFs) that are organic–inorganic composite materials with large surface areas, adjustable pore structures, and good thermal and chemical resilience and stability. Unfortunately, they are limited by their fine powder structure, making them difficult to use in industry. Significant efforts are currently underway, with expectations of new advancements in the development of these innovative membrane materials.

The CO_2 membrane separation process includes cooling of the flue gases (absorber) to an operating temperature of the membrane material. There are different types of membrane plate-and-frame design concepts, with most common types being hollow fiber, flat sheet and spiral wound (Basile et al. 2010). In addition, for CO_2 separation there are two types of membrane technologies: gas separation membranes and gas absorption membranes. In the gas separation approach, a higher flue gas pressure is applied to one side of the membrane, and the other is kept at lower pressure. For absorption membranes, a microporous membrane structure (typically in a hollow fiber design) is used with flue gas flow on one side and absorbent liquid flow on the other side.

Chemical looping combustion (CLC) is a technology that has high potential for CCS and is economically feasible. It uses cyclical processes and a set of sub-reactions by utilizing looping materials (LMs). This process can be used to create a combination of heat, electricity, fuels, and chemicals (Asif et al. 2018). CLC process includes two reactors. One is used as an air reactor with a purpose of using an oxygen carrier—metal particles (e.g., copper, iron) and oxidizing them using air. These metal oxide particles are then transported to the other vessel, fuel reactor. Here, the metal oxides react with the fuel used and produce energy and flue gas streams of H_2 and CO_2, from which CO_2 can be derived. The looping materials are then regenerated in the air reactor, thus creating a loop. A derivative of the CLC process is calcium looping (CaL) process. It is founded on a principle of carbonation-calcination equilibrium reaction. It is a reversible reaction involving calcium oxide and CO_2. Following a similar approach as with CLC in the initial reactor, calcium oxide (CaO) and carbon dioxide (CO_2) are combined through a process known as carbonation. The resulting product, calcium carbonate ($CaCO_3$), is then transported to the second vessel, the calciner reactor. Here, fuel is burned using oxygen atmosphere to create heat for a reversed reaction producing high purity CO_2. Cyclones are used in the transport flows for solids separation. This technology (CaL) as well as the CLC use well established and mature circulating fluidized bed (CFB) technology concepts and therefore is of low risk and scalable. Also, as the absorbent is lime and as such very low cost the process is seen highly economic for industrial applications.

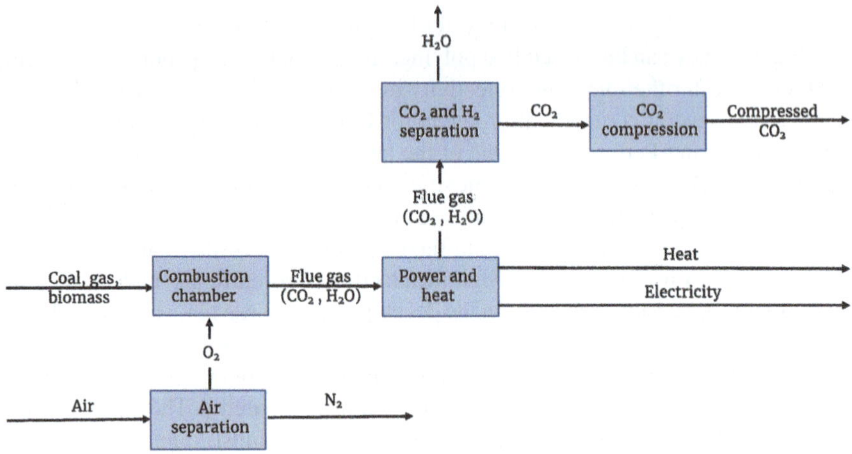

Fig. 5.6 Oxy-combustion CO_2 capture method block diagram for heat and electricity generation

5.4.4 Oxyfuel Combustion

Oxyfuel combustion CO_2 capture is considered as a highly potential technology in CCS. The basic approach is to use oxygen in the combustion process instead of air. This is done by separating oxygen from air by removing almost all nitrogen (air separation unit). Unfortunately, it also is creating a challenge of keeping the cost competitive compared to other CCS technologies. The flue gas in this technology is inherently richer in CO_2 than with traditional air combustion. In addition, removal of nitrogen reduces NOx emissions, and the higher CO_2 concentration allows an easier removal of the pollutants such as sulfur oxide (SOx) and mercury, particulates from the flue gas. The final step is to use gas processing unit (GPU) to treat and process the downstream flue gas for the desired purity of CO_2 and for its transport and storage (Fig. 5.6).

5.5 Industry 4.0 Technologies for Carbon Neutrality

5.5.1 Basics

Since the beginning of the industrialization, the evolution of industrial manufacturing technologies has progressed from mechanization, water, and steam power to electrified mass production and assembly lines. After these two evolutionary steps, a so-called third wave emerged and included embedding electronics, automation, and Information and Communications Technology (ICT) systems into the manufacturing processes. The most recent evolution has been the emergence of Cyber

Physical Systems (CPS) that include the design and integration of a network of physical devices and objects that interact as a system. This approach brings the benefits of efficiency, autonomy, adaptability, reliability, and safety and includes various technological advancements in computing, control systems, networking, and sensing.

Some of the more specific technologies or platforms that are considered part of the so-called industry 4.0 technologies (industrial evolution towards automation and data exchange) include *Internet of Things* (IoT), *Artificial Intelligence* (AI), *and Cognitive Computing*. These technologies enabled also major developments in industrial ecosystems and infrastructure. The industry 4.0 therefore includes not just the manufacturers but also other key members of the value chain. These consist of the entire supply chain and various service providers, and actors such as cities and communities including start-ups, SMEs and large global corporations, research institutions, and academia.

The impact of industry 4.0 to climate change mitigation and carbon zero efforts globally can be extremely high. As industrial manufacturers contribute roughly 30% to carbon emissions (823 million metric tons of CO_2 emissions, 2019) (Jovaid et al. 2022), the manufacturing industry needs to continuously make strides in reducing their emissions. In order for the industry to reach or closely approach carbon–neutral manufacturing status, it must make changes in its industrial ecosystem and the way business models, and manufacturing processes are designed. As the world's energy landscape is changing and becomes increasingly geopolitical, the fact that the manufacturing industry uses about 54% of the total world's energy delivered, efficiencies, and energy strategies have become progressively more important. In this context, the national security and energy independence have already become a top priority (Shepard and Pratson 2022).

The industry 4.0 specifically helps reduce supply chain CO_2 emissions that are typically more than 5 times higher than actual industrial manufacturing operations create (Zhang et al. 2022). The new automated and connected processes in the manufacturing industry can reduce and manage the complexity of these global supply chains. Connecting partners and stakeholders with more transparent data and sharing it also with consumers will enable a more responsible and accurate supply chain-wide management of the industrial carbon footprint. This ultimately helps them identify the systemic opportunities to reduce the CO_2 footprint. Learnings from this collaboration will eventually inform and drive further actions collectively including the development of smart supply chains using industry 4.0 technologies. Overall, the benefits of industry 4.0 approach to manufacturing create opportunities to design and manufacture greener products, further advance the manufacturing processes by increasing efficiencies and reducing waste and use of natural resources and most importantly developing the industrial ecosystems towards carbon neutrality.

5.6 Internet of Things (IoT)

IoT in the industrial setting is a network of connected physical objects (e.g., machinery, valves, control systems) that have the ability to collect data using various types of embedded sensors and exchange it in real-time across the Internet-based communication platforms. Expanding this to beyond industrial setting the number of connected devices is estimated to be already 30 billion is 2020 (Munirathinam 2020). In the world of industry 4.0 or smart manufacturing, this technology enables to collect relevant data from the whole life cycle of the material used in the manufacturing of the product. This cradle-to-grave data collection can better enable new business models such as circular economy to be implemented in an economically attractive way.

Keeping the raw materials, components, and parts use in a circular model, and tracking the ones with high economic value will enable to dramatically reduce the use of raw materials, water, chemicals, and energy. IoT will assist in creating cross-industry and value chain-wide approaches to product innovation. For example, a waste or product (e.g., packaging) from one part of the value chain may be a useful raw material or product in some other part as it moves across the supply chain. These developments will take time as the manufacturing industries do not have the history of such interconnecting ways of working nor data collection, processing, and handling. Other technologies such as cloud computing, signal processing, and AI will further facilitate this transformation process for the manufacturing industry.

IoT is based on four key concepts: hardware or device, connectivity, the data that is collected, and data analytics for processing the data to a useful form. One of the components in the IoT is the radio frequency identification (RFID) tags (including radio transponder, receiver, and transmitter) which allows the device or machine or computer to recognize relevant items, to record data or data that provides information about other data (metadata) and also operate or monitor other devices using radio waves. When the RFID reader is connected to the Internet, it can be used to monitor and operate other devices embedded with RFID tags globally, automatically, and in real time, if needed. RFIDs are good at transmitting smaller amounts of data from an RFID tag to a reader, however, within a short distance (depending on type and the frequency used from 1 m up to about 15 m). Major improvements have already been achieved in real-time tracking of business inventories and assets obtaining major improvements in operational efficiencies, reducing energy and waste in for example retail industry and logistics.

Passive RFID equipped sensors can also be used in manufacturing processes where physical connectivity is not feasible. It also allows new business model concept in retail such unmanned stores, self-checkout systems, and smart mirrors for apparel business. The RFID-based electronic shelf labels are becoming more popular and further reduces the use of shelf packaging and the implementation of self-checkout systems and dynamic pricing concepts. Another set of versatile technologies for connectivity includes Wi-Fi, power-efficient Wi-Fi 6, and the more recent Low Power Wide Area Networks (LPWANs). LPWANs, in particular, are specifically designed for large-scale IoT networks in both industrial and commercial buildings. They operate in low bit rate mode and can connect to battery operated sensors.

Industrial facilities use both of these technologies for various device and machinery-related connections. Smart industrial buildings where Wi-Fi connected camera, security, surveillance, and monitoring systems are improving overall availability of the industrial processes are some of the examples. Most all smart devices and appliances are also in industrial facilities Wi-Fi connected including smart lighting sensors with automatic turning on and off the lighting creating further efficiencies and energy savings. In terms of heating and cooling efficiencies thermostats in industrial operations are connected and also using sensors the maintenance of air ducts, filters and HVAC systems can be made smarter and more accurate. The drawback in especially using Wi-Fi technology remains to be coverage, scalability, and power consumption. Zigbee is older shorter-range wireless technology protocol that uses relaying approach to extend sensor data coverage over multiple sensor nodes in industrial facilities. Other technologies for connectivity in industry include Bluetooth or Bluetooth Low Energy (BLE), with Bluetooth being able to transmit high quantities of data with high-energy use and cost. BLE is designed for applications exchanging less data and with low-energy consumption leading to long battery life and overall cost. Cellular networks (e.g., 3G/4G/5G) are not necessarily practical for majority of IoT applications; however, they fit well in specific industrial uses including connected cars or fleet management in logistics. Some additional examples include traffic routing and tracking services and advanced driver assistance systems (ADAS) for autonomous vehicles that can be used in manufacturing sites. The onset of 5G provides high-speed and ultra-low latency (ultra-low tolerance for delay). It can also improve the real-time industrial monitoring and surveillance and in medical industry and highly time-sensitive industrial automation and sensory applications.

In summary, IoT offers several benefits and applications for advancing carbon neutrality. It facilitates operational support through real-time production and operations performance optimization, encompassing reporting, issue resolution, and controls. Maintenance is another major benefit adding to productivity and energy savings and includes reliability improvements and availability of industrial processes using data analytics and digital process support. Energy efficiency benefits come from process and automation optimization and issue resolution in real time. Finally, Environment, Health, and Safety (EHS) benefits are also one important advantage and come from root cause analysis and preventive maintenance and automated monitoring and documentation.

5.7 Artificial Intelligence (AI) and Cognitive Computing

AI is recognized as one of the leading technologies shaping the future of mankind. What AI really represents comes down to what intelligence means in the context of engineering, science, and business. It is in its core the ability to attain knowledge and to use it to accomplish an outcome. Now, the principle of AI systems is to take in vast amounts of data and to analyze it so that patterns and correlations can be found and used to create projections of future states. The official simple definition is challenging and exists in various forms based on applications (Copeland 2023). The distinction is however generally agreed to include the ability for machines and computer systems to solve problems, demonstrate rational behavior in complex problem solving, and take appropriate actions to achieve the goals that come across or are presented to the AI.

The history of the concept of AI can be traced back to the times of Aristotle in Greece (384–322 BC), when the intellectual lexicon, and methods of inquiry and problem solving was developed and advanced to a level that the progress of our society still is benefiting from. The next major milestone could be credited to the first design for a programmable machine by Charles Babbage and Augusta Ada Byron, Countess of Lovelace (1836). Another major step was the first mathematical model of a brain's neural network by Warren McCulloch and Walter Pitts (1943) and development of the imitation game (later called The Turing test) by Alan Turing (1950). This test was designed to evaluate a computer's ability to demonstrate intelligent behavior that cannot be distinguished from human. The term and concept of AI can be credited to John McCarthy was an AI pioneer who also invented the LISP computer language that is an AI language that continues to be used to this day. After a period of relatively low support for AI development by both the public and private sectors, investments in AI began to rise again in the 1980s. This resurgence was marked by the introduction of "deep learning" by John Hopfield and David Rumelhart, which enabled computers to learn through experience. Additionally, Edward Feigenbaum introduced decision-making abilities into AI during this era. Some of the publicly recognized highlights in the AI development includes IBM's Deep Blue computer's defeat of Chess Grand Master Gary Kasparov (1997). Advancements in speech recognition and the use of Big Data for advanced data analytics continue to make major strives, and Google's AlphaGo program was able to defeat defeated a South Korean master of Go, Lee Se-dol, in an ancient board game of high complexity (2016). In 2017, a program is designed by Tuomas Sandholm, called Libratus, and was able to win top human poker players (2017).

In principle, AI can be grouped into two categories based on functionality and capabilities. In terms of functionality, there are four sub-categories. *Reactive Machine* is the first, and in this category, the system has no memory power and does not have the ability to learn from past actions. The second category is *Limited Theory*, which has some but limited memory, and uses it to improve its decision making. The third category is *Theory of Mind*, in which is under development and is targeting at reaching true rational decision-making capabilities similar to humans. The last category is *Self-Aware* and is more a vision or hypothesis at this point aiming at AI to have self-awareness of their internal emotions and mental states. The capabilities of AI can also be sub-categorized and include *Artificial Narrow Intelligence* (ANI), which means that the AI is performing well defined and quite specific programmed functions and tasks with limited memory. Most of the AI applications are currently in this category. *Artificial General Intelligence* (AGI) is more advanced and able to train, learn, and understand somewhat similar way as a human. This is still a work in progress, just like the Theory of Mind. Finally, *Artificial Super Intelligence* (ASI) would be able to outperform humans in various tasks and due to its computing, memory, and decision-making capabilities. This is again a conceptual vision that the AI filed is predicting to be reached eventually.

The core of AI systems is powered by algorithms that use techniques machine learning and deep learning techniques. Machine learning is the computer's ability to simulate and mimic the human reasoning to predict and react to its environment by the use of new information learning and problem solving. The deep learning is another higher level of machine learning and uses multi-layer (three or more layers) artificial neural networks for the simulation of the actual behavior of the human brain. In practice this means adding more "layers" to learn from the data, thus creating deeper insight to data analytics and allowing it to "learn" more from large data sets helping it to optimize and refine the accuracy of the predictions and actions it takes.

The advancement of AI is exponential and major progress has been made during the past 5 years in several application areas including image recognition, the ability to identify objects, people, places, human writing and dynamic changes, and actions from video and images with almost limitless capacity. The integration of image recognition and motor control for robotics and automated systems has also advanced. Speech recognition and generation have reached the next level and are now used in most mainstream communications technology applications. Furthermore, natural language processing, which encompasses understanding and generation of human language, has evolved from a work in progress to practical applications. These applications now include language translation and the availability of virtual personal assistants. With advancements in decision making and planning process capabilities, the applications have also expanded to not just include games but also medical diagnosis support, advanced logistics systems, and EV autonomous driving.

As the terminology AI implies that computers or machines are going to think for themselves, the term *Cognitive Computing* instead is purposed to create machines that are able to analyze tremendous amounts of unstructured information (vibration patterns, images, speech). Using such analytical tool, Cognitive Computing can recognize and interpret hidden patterns in data and construct meaning to support decision making.

Manufacturing companies have invested in digitalization of their production plants and have established distributed and supervisory control systems. The visualization of process controls for operators has greatly advanced, and predictive maintenance analytics has been developed and implemented. However, the latest advancements of AI and Cognitive Computing including decision-support solutions, advanced analytics and machine learning have not necessarily reached all manufacturers. Still many factory control room operators rely on experience, intuition, and judgment. With AI, operators can change from visual set-point information usage to a more autonomous mode. This way the fine-tuning for optimized performance is given to AI, and more complex tasks can be left to the human operators. The AI system's advanced analytics can improve the energy efficiency, production yield (percentage of non-defective items), and throughput time (product pass through time) of heavy-asset production operations. The advanced video and image analytics and surveillance can also automatically analyze production events, uncover anomalies, and obtain operational insights.

In the future, manufacturing will have more robots that are able to work in a collaborative way in warehouses and production lines. Machine and deep learning will improve the predictive maintenance accuracy and decision-making processes. Products will be increasingly manufactured to order and to the customer specifications and demand when AI and Cognitive Computing are able to predict the purchasing habits. Real-time optimizers will be used for company's core manufacturing assets creating efficiency increase. All this computes to reduction in emissions per production ton and thus GHG emissions from both industry and services sectors.

5.8 Carbon-Free and Renewable Energy

5.8.1 *Basics*

When we look at the concepts of *carbon-free energy and renewable energy*, it is important to understand the underlying difference between them. Both are used in similar context; however, carbon emissions, environmental issues, and economic impacts can be quite different. The meaning of carbon-free energy is straight forward and states that no carbon emissions are being produced from the energy production. For example, hydropower is considered as renewable energy, since it relies on the water cycle, which is caused by the sun, making it renewable. Environmental implications may, however, still exist. For example, as building a large dam system to secure hydropower most likely will have major negative biodiversity implications. Nuclear power on the other hand is carbon free but not renewable. The fuel is uranium, and it is processed into small ceramic pellets that are stacked into fuel rods and used as the reactor fuel.

5.8.2 *Wind Power*

Wind power has always been connected to agricultural and industrial development. It converts the kinetic energy of wind into mechanical power. This has been then used more traditionally for applications including sawmills, grinding grain, or pumping water. Modern wind power includes a turbine that converts the movement of the blades into electricity by a generator. There are two types of turbines: *horizontal-axis wind turbines* (axis parallel with the ground) are the most known type and typically have two to three blades and operate with blades facing into the wind. The other currently less common type is *vertical-axis wind turbine* (the rotor shaft is transverse to the wind). They are omnidirectional, which means that they can operate without turning them towards the wind. Wind turbines can be land or offshore-based, and they produce power ranges from hundreds of kilowatts to a few megawatts. These systems are many times built as a group, and it is preferred to build larger mills for increased cost efficiency. Offshore wind power systems are larger in size with wind blades normally ranging from 150 to 250 m and height from 100 to 150 m. Their power generated can range from a few megawatts to a few tens of megawatts. They benefit from higher wind speed and greater wind consistency compared to on shore wind turbines. They also benefit from the installation-related logistics due to more easy transportation to their location by ships.

Although wind power is sustainable, creates jobs, and can benefit rural areas economy, it also has weaknesses as well. It occurs at irregular intervals; it is still not quite competitive from economic perspective; best installation locations are far from the locations of use (cities or manufacturing cites) and therefore transmission lines are needed, creating additional capital investment needs and higher transmission energy losses. Wind turbines installations are also destructive to landscapes and create noise pollution. In addition, it negatively impacts local wildlife and reduces biodiversity. The cost of wind turbines has come down during the year but remains at about \$1.3 million per megawatt (MW) and operating and maintenance costs are roughly 20–25% of the cost per kWh produced over the life of the turbine, and they increase by the age of the system (EWEA 2009). The wind turbines operate on average with a capacity of 35–65% due to the constant change in wind conditions. The main drivers in profitability of wind turbine farms include revenues from sales of electricity, costs of capital expenses (CAPEX), and operating expenses (OPEX). As the operating costs are rising the operators are putting more focus on the procurement and logistics for components and the infrastructure for grid connection. In additional operational efficiencies and reduced maintenance costs are being addressed by developing AI-based predictive maintenance, and extending the wind turbine lifetime, by predictive maintenance and optimized service contracts.

The worldwide onshore wind electricity generation was 1480 TWh and the offshore 112 TWh in 2020 (IEA 2022). To reach the global net zero emissions targets of 2050, an additional 8 000 TWh of green total capacity is required in 2030. This means an average wind power capacity increase of 18% per year is needed during 2021–2030 (IEA 2022). The current leadership in the installed capacity is in China and the USA (79% of total combined capacity). Enhanced global endeavors are necessary to attain continuous capacity growth, alongside technological advancements aimed at reducing costs, promoting component recycling, and improving overall efficiency. In additions, the offshore wind utilization should be expanded and onshore wind permitting should be more supported.

5.8.3 Solar Power

Solar power is based on conversion of energy from sunlight into electricity or heat. Most straightforward approach is by directly using *photovoltaics* (PV) for electricity generation, but also indirect approaches can be deployed by using *concentrated solar power* (CSP), or hybrid approaches (combined electricity and heat). The first PV dates to 1883 when inventor Charles Fritts developed the first solar cell using a thin gold coating over semiconductor material selenium. The principle is the same as today; however, the efficiencies were at a one percent level compared to today's PVs that have efficiencies between 15 and 20%. The most efficient PV or solar panels are currently manufactured by Sunpower (USA), LG Solar (South Korea), and REC Group (Canada). However, the largest global manufacturers are LONGi Solar, Tongwei Solar, and JA Solar. These suppliers are all from China, which holds a market share of more than 80% for all solar panels. China also accounts for more than double of global demand; thus, China had $28 billion worth of solar cell exports in 2021.

The CSP system uses a different approach to power generation. It uses lenses or mirrors and specific tracking systems to concentrate sunlight as efficiently as possible. The resulted thermal energy is then used to generate electricity by deploying conventional steam turbines. There are three main approaches to the CSP: *Linear Concentrator Systems* (LCS), *Dish Systems* (DS), and *Power Tower Systems* (PTS). In the LCS, long parabolic or Fresnel lens-based collectors are used to concentrate sunlight onto receiving tubes. These tubes have a heat collecting fluid that is used for boiling water and creating steam for a turbine generator for electricity generation. The DS are using a slightly different approach by having large, mirrored dish typically made of number of small mirrors that are organized into a satellite dish shape. Using tracking system, the dish follows the sun and enables the concentrated sunlight to reach the thermal receiver, absorbing and collecting the heat and transferring it using heated fluid to for example Sterling type heat engine and eventually for electricity generation. The PTS on the other hand has large, flat mirrors (heliostats) and by using a tracking system it concentrates the sunlight onto a top of a tower, where the receiver is positioned. Similarly, to LCS heat fluid (or sometimes nitrate salt) is used to collect and transfer the heat to be used in a steam turbine. These systems are also capable to store thermal energy for electricity generation when sunlight is not available. *Hybrid technologies* are used to turn solar radiation into both thermal and electrical energy. These systems may combine regular or concentrated (using lenses or mirrors for focusing sunlight) PV panels and solar thermal collectors in a paired approach. Or some systems are combining PV and thermoelectric devices that use temperature difference between dissimilar materials into an electric current.

The global solar power capacity has increased tremendously during the past decade. Only in 2021, 168 GW capacity of new solar PV was installed. In 2022, it has reached approximately one Terawatt capacity and is forecasted to expand to a level of 2.3 TW in 2025. Globally, PV power accounts for about 3% of electricity generation. It is also *the third-largest carbon-free electricity producing technology*.

PV power has already passed the bioenergy in 2019 and represents the second-largest growth in all renewable technologies, slightly behind wind and ahead of hydropower. However, much more effort is needed as the set global net zero emission target for 2050 demands a fivefold increase in annual capacity until 2030 (current capacity is about 900 TWh and target for 2030 is 6970 TWh). This includes both regulatory and financial challenges and calls for more public and private efforts in policy commitments and special support and guidance for emerging and developing countries.

5.8.4 Nuclear Power

Based on 2021 estimates, nuclear power accounts for estimated 10% of global electricity production. The number is even higher for the industrial OECD nations and close to 20%. Compared to other carbon-free sources, nuclear power meets large-scale demands that the growing electrification is creating; need for electricity will rise by more than half to 2040 (IEA 2019). Nuclear power is attractive option due to its production reliability and predictability that are built within to the technologically advanced engineering and highly regulated operational aspect it has.

The main idea of power generation for nuclear power is based on the energy release coming from splitting the atoms of nuclear fuel that typically is low-enriched uranium. This generates nuclear fission and that is then engineered in nuclear power plants to be sustainable for continuous power generation. The fuel rods have compacted uranium pellet that weigh about 10 g each (a rod may contain some 10 million pellets), containing energy that amounts to a ton of coal or 481 m^3 of natural gas. The fuel pellet provides long-term (up to five years) heat for power generation. Uranium is also globally abundant element; thus, it can provide energy for nuclear power plants for generations to come. The process was discovered and developed in the 1940s. It started first as weapons research and then continued to civil use of nuclear fission. Undisputedly, nuclear power, during its operation, does not create any CO_2 emissions. Interestingly, already now one-third of the world's carbon-free electricity is produced using nuclear technology, and thus, it is becoming increasingly important in meeting climate change goals.

The classification of nuclear reactors can be done using various criteria, such as *moderator material (graphite, water, light element), nuclear reaction (thermal-neutron, fast neutron), coolant (e.g., pressurized, supercritical, boiling, and heavy water, liquid metal, molten salt, and gas cooled), and use (e.g., electricity generation, heat generation, propulsion)*. The majority of world's nuclear power plants (more than 80%) are so-called *pressurized water reactors* (PWR). In this reactor type, a pressure vessel is used that comprises of nuclear fuel and control rods, coolant, and moderator (purposed to slow down high-velocity neutrons for higher fission efficiency). The coolant water in the pressure vessel is run through a steam generator, producing steam in the secondary (non-radioactive) loop of water. This steam then runs the turbine system for electricity production. A sub-type is to use heavy water (D_2O) as a coolant and thus avoiding the use of graphite moderators, poising radiological, and dust hazards. Another PWR-related reactor type is *Boiling Water Reactor* (BWR), which is not using a steam generation loop. It is operated in a lower pressure mode, and therefore, the cooling water boils in the pressure vessel and produces the necessary steam for turbines. The design is simpler and is considered more durable and secure. A *Pressurized Heavy Water Reactor* (PHWR) is similar as the PWR but using heavy water (is a form of water that contains only deuterium) as the coolant. It has several advantages including ability to use natural uranium and to increase neutron efficiency. The drawback is the high cost of heavy water. The design is also known as Canadian design or CANDU. Lastly, a gas-cooled reactor (GCR) and advanced gas-cooled reactor (AGR) designs are mostly used in the UK and use graphite as a neutron moderator and a gas (CO_2 or helium) as a thermally higher efficiency coolant for the heat transfer to the steam generator. One drawback is the large volumetric size needed for the reactor core.

A design that does not use moderator media *is liquid metal fast-breeder reactor* (LMFBR). It is a fast neutron system, and due to its nature of neutron capture, it also produces fissionable fuel more than it consumes. This design is stable as it does not demand high-pressure containment due to the use of liquid metals such as lead and sodium, for the coolant. There are pros and cons using lead as it is transparent to neutrons and thus creates high neutron efficiencies and it does not get radioactive. It is also inert with lower explosion risks compared to sodium. However, the drawback is that it is toxic, and its eventual disposal can be very expensive. Some of these flaws are mitigated using a lead–bismuth as the coolant. Sodium is lower cost coolant than lead and additionally has corrosion preventing properties. A shortcoming is that it is highly explosive when it is exposed to water; thus, a possible sodium leak can be a major safety issue.

Another interesting reactor type is *molten-salt reactors* (MSR) that use fluid nuclear fuel (nuclear fuel dissolved in high temperature fluoride or chloride salt). This way the fuel salt can be used as the heat producing fuel and the coolant that transports heat to the steam generator and turbines. This design has the benefit of not having the need to use higher pressure coolants nor any explosive or flammable substances in the reactor. It also has high efficiency and power density, making it attractive to mobile systems. This type reactors have been designed potentially for aircraft propulsion purposes. Somewhat similar types of *aqueous homogeneous reactors* (AHR) have also been designed using soluble nuclear salts such as uranium sulfate or nitrate.

A more novel approach for reactor design is *pebble-bed reactor* (PBR), which is a very high temperature reactor with graphite moderator and gas cooling. This reactor type was proposed already in the 1940s by Farrington Daniels from USA. Its principle is to use spherical or cylindrical fuel elements composed of graphite and also contain microfuel particles. These particles (also called TRISO particles) have a core of uranium oxide (UO_X) and have four other layers. These include porous layer of carbon, layer of protective pyrolytic carbon (PyC), ceramic layer of silicon carbide (SiC), and lastly a dense external layer of PyC. These layers are deposited using fluidized chemical vapor deposition (FCVD) technique and create a controlled containment of fission products and protective fuel structure. The pebbles are able to handle much higher temperatures than more traditional reactor designs. Thousands of pellets are used in the reactor, and cooling is accomplished by using CO_2, helium, or nitrogen gases. This reactor type is considered safe (less radioactive contamination in the core) and provides high thermal efficiencies due to its high operating temperature. As this reactor type can handle high temperatures, it can also cool by natural circulation and therefore survive in most accident scenarios. Due to its design, its operational high temperature levels give also higher thermal efficiencies compared to more traditional nuclear power plants (up to 50%). There are several new reactor concepts that are in the design phase and are to be tested and commercialized for various applications. These reactors are called generation IV or V + reactors. Some of them are more or less in a research phase and therefore would not be feasible for commercial construction before 2030, perhaps by 2050. The main goal for this development is the advancement of nuclear safety, waste reductions, and efficiencies. Such concepts include for example *hydrogen-moderated self-regulating nuclear power module* (Peterson et al. 2008) and *small modular reactors* (*SMRs*) (Tulkki et al. 2017).

Clearly, the uranium waste disposal is a broader nuclear energy issue as there are over 400 nuclear power plants globally with more than 11,000 tons of spent fuel per year. Just to clarify the environmental and economic implications; *Low-Level radioactive Waste* (LLW) is compacted in steel canisters and stored in concrete vaults at ground level, or in caverns at depths of tens of meters. The *High-Level radioactive Waste* (HLW) or *Intermediate-Level radioactive Waste* (ILW) containing long-lived radioisotopes is stored in deep geological disposal locations at depths of a few hundred meters for mined repositories, or thousands of meters for boreholes. It should be stated that actual carbon free or low carbon does not exist as there is always some CO_2 emissions from the infrastructure, logistics or operations.

As stated earlier, nuclear power is critical in achieving net zero goals. The policies in several OECD countries have not been supportive of nuclear energy, and many reactors have been permanently shut down. For example, the net gain in capacity was only 0.6 GW due to the shutdowns. The estimated nuclear power capacity will be 582 GW in 2040. This is about 150 GW less that the net zero emissions target for 2050 would demand. Unless the annual capacity increase is doubled, we will not be able to keep up with the 2050 scenario goals. Again, financial incentives for further investments and policy changes such as the latest European Commission declaration of nuclear power to be green are urgently needed in order to gain necessary traction to growing nuclear power capacity globally.

5.8.5 *Hydropower*

One of the oldest renewable energy sources is hydropower. It is used to generate electrical power from natural flow of water by using a dam or other diversion structures. The system is based on a continuously renewing water cycle, and hydropower plants are located near rivers or other water source. The electricity generated is dependent on the *captured water flow and the elevation or fall* (*head*). The more flow and head the more electricity are created in the turbines. Hydropower is considered large when the capacity reaches to a level higher than 30 megawatts (MW), and small hydropower plants are typically in capacity somewhere in between 100 kilowatts and 10 MW. There are four basic types of hydropower systems including *run-of-river facility, impoundment facility, pumped storage hydropower,* and *offshore hydropower.*

A *run-of-river facility* or also called diversion operates by taking a portion of a river and channeling it through a canal, gate, or intake structure that controls water flow. The riverbed elevation is used as the head to produce electricity by flow of water to turbines. This approach does not necessarily demand the use of a dam. An impoundment facility is larger in size and mostly uses a dam structure for water storage. This way the water flow can be regulated and discharged based on demand and other factors from the reservoir to generate electricity. *Pumped storage hydropower* functions as a physical energy reserve, stockpiling energy that can be produced by other perhaps less consistent energy production means such as wind or solar. The energy is stored by pumping water from a lower-level reservoir to a higher level using this electricity that is needed to be stored. Eventually based on demand the water is discharged back to the lower reservoir through a turbine. *Offshore hydropower* uses wave motion or tidal currents for electricity generation from seawater. For tidal currents, a dam-like structure with turbines placed at the bottom of the reservoir is called tidal barrage and is used to separate the sea and the tidal basin. It works similarly to any large hydropower reservoir dam, releasing water in the through turbines and generating electricity.

The global capacity for hydropower in 2020 was 4418 TWh. It accounted for more than other renewable energy production combined. Looking at the net zero targets, there should be added hydropower capacity growth of 3% per year. This would lead to 2030 milestone of 5870 TWh and keeping the 2050 target intact (IEA 2021).

5.8.6 Biofuels

The history of using biofuels is long and traces back to times before the development of human civilization. Since the discovery of firewood, it has been used to cook food, and heat shelter and housing. Wood has been used to run trains, produce electricity and when the early designs of diesel engine and the beginning of the first serial production cars, biofuels have been considered as an energy source. The advent of low-cost fossil fuel such as oil, coal, and gas diminished the use of biofuels for transportation, power generation or industrial use. The Second World War created a new strategic need for biofuels due to fossil fuel shortages and the rapid advancement of science and technology created concepts such as bioethanol mix with petroleum. Also, due to the shortage of petroleum, Fischer–Tropsch process, an advanced biofuel conversion technology that is used to synthesize liquid (or gaseous) biofuels, was advanced and commercialized by Germany during World War II. It was used to produce fuel for German automotive fleet during the war (replacing 25% of fuel needs at that time). Eventually, Fischer–Tropsch production accounted for about 10% of German war production of fuels (Rahardjo 2013).

The biofuels can be categorized based on their production technology and sources of biomass and are called: first-, second-, and third-generation biofuels. The first-generation biofuels include ethanol and biodiesel and are based on using mostly edible crops with sugar, starch, or oil content. This biomass is converted into ethanol or biodiesel, using well established technologies in fermentation of sugars (C6 sugars, or hexoses such as glucose) or by transesterification. In practice, mainly corn, sugar-cane, and sugar beets are used for the ethanol and soy, canola, and palm for biodiesel production. The global trend in the increased use of biofuels has been driven by several factors including the volatility of oil prices, the need for actions regarding climate change mitigation, and also the need for economic support for the rural areas.

Due to these reasons, the production of the first-generation biofuels increased tremendously (about 10x) after the implementation of US Renewable Fuel Standard (RFS), an Energy Policy Act of 2005 and the following Energy Independence and Security Act of 2007 (EISA). The first-generation biofuels have always been a disputed topic with controversy. Further, the US Environmental Protection Agency (EPA) drafted minimum biofuel blending policies to transportation fuels similarly as EU by requiring 10% blending of biofuels. In general, the first-generation biofuels have shown to be commercially viable option; however, there are several drawbacks. One major issue is the fact that they are directly competing with food and thus drive global challenge of food scarcity, driving food prices high and reducing the supply. The other issue is the actual performance in terms of mitigating climate change. Its sustainability aspects on impact to agriculture and land use patterns are questionable. For years scientists have viewed biofuels carbon neutral and have assumed that plants used for biofuels feedstock absorb CO_2 during their growth and effectively offset the CO_2 emitted in the consumption of the biofuels. The recent LCA studies have shown that the GHG emissions induced by increased biofuel consumption are controversial and uncertain (Løkke et al. 2021). The scientific estimates seem to vary with wide margins, and there is debate over the environmental impact on various factor in cultivation, deforestation, and soil acidification. The conclusion seems to be aligning towards first generation biofuels being source of much more GHG emissions than expected. One of the issues that has been discarded in the estimates has been Indirect Land Use Change (ILUC) emissions. It occurs when forest and grassland are transformed to cropland for biofuels feedstock production. The new results point that ILUC-related GHG emissions negate the climate benefits otherwise estimated for some of the biofuels, especially when it comes down to first generation.

The second-generation biofuels are based on more advanced technology and produced from non-food biomass that comes from a wider range of different feedstock, ranging from lignocellulosic feedstocks from trees to waste including corn stover, wheat straw, forest residue and municipal solid waste. The value of these biomass is much lower than the value for first-generation biofuel feedstock, which also establishes an economic incentive. However, the production technologies are more expensive and still not fully mature. New concepts including "biorefinery" have emerged from the second-generation approach to biofuels. In this approach, biomass used will be "refined" in a comprehensive way to produce multiple products. This further enables such business models as circular economy or circular bioeconomy.

The main technological approaches to the second-generation biofuels include *thermochemical and biochemical conversion processes.* The thermochemical processes include gasification, pyrolysis, torrefaction, and hydrothermal liquefaction. All these processes produce such gases as hydrogen, carbon monoxide, CO_2, methane and other hydrocarbons, and water. Pyrolysis is performed in temperatures ranging from 400 to 800 °C and creates tar, bio-oil, and char. Torrefaction uses temperature between 200 and 300 °C and converts biomass feedstock to a more easily transported and stored form with higher energy value and yield for further conversion. The gasification gas is created in temperatures typically > 700 °C and is called "synthesis gas" or "syngas" and can further be converted to different types of fuels, such as ethanol, synthetic diesel, synthetic gasoline, or jet fuel.

Hydrothermal liquefaction is also called "hydrous pyrolysis" and has the benefit of being able to process wet biomass. The process temperature is mostly at temperatures ranging from 280 up to 400 °C and higher pressures (10–25 MPa). One major benefit for this technique is its flexibility in using a wide range of feedstock materials. The biochemical conversion process uses different enzymes, different bacteria, and other microorganisms for breaking down biomass into gaseous (biogas) or liquid fuels (bioethanol) (Balat 2011). This can happen via different processes such as anaerobic digestion, fermentation, or composting. In terms of lignocellulosic feedstock, the process demands several pretreatment steps due to the fact that cellulose as well as hemicellulose in feedstock cannot be processed easily with microorganisms. The material needs to be refined so that the surface area can be higher, which then increases the accessibility for hydrolysis reaction. This can be done using different physical, chemical, and biological pretreatment methods. The next step is to use hydrolysis (e.g., acid or enzymatic hydrolysis) to separate out lignin, hemicellulose and cellulose and to use cellulosic fractions: cellulose and hemicellulose, for creating sugar monomers (C6 and C5) for fermentation to alcohols. In case of municipal waste or landfills, fermentation techniques are used for biogas production and include specific and also genetically modified bacteria.

In general, there are three use categories for biofuels that are produced from biomass by technologies described earlier (using thermal, biological, or chemical processes). *Biodiesel* is one of the three and for example in the USA it is the second most used biofuel (bioethanol being the largest). The production capacity in the USA is some 75 biodiesel production plants and a total of 2.4 billion-gallon production capacity. In addition, 200 million gallons are imported. This fuel type is not an actual "drop-in" fuel that has the same chemical composition as petroleum diesel. This raw biodiesel is limited in its use and transportation as it functions as a solvent for fuel lines, pipelines, and tanks. Biodiesel can have detrimental effect to transportation vessels and engines using it. So-called *renewable biodiesel* (or *green diesel*) is made from the same oils and fats (e.g., vegetable oils, animal fats, used cooking oil) as the biodiesel but instead of using transesterification (using alcohol to process feedstock to form methyl esters) it is processed by hydrotreating (removal of oxygen, nitrogen, sulfur) or other biochemical and thermochemical technologies. *Renewable biodiesel* is so-called *drop-in fuel* due to its similar chemical structure as petroleum diesel (R100). It can be used as is or as a blend with petroleum diesel. It also does not have any detrimental effect to diesel engines, nor transportation vessels. The capacity for the production of this fuel type in the USA is much smaller than for biodiesel as it demands more complicated and capital expensive processing. Due to its legislation, California uses most of US renewable diesel fuel imports. Another type of biofuels is *Sustainable Aviation Fuel (SAF), alternative jet fuel (AJF) or biojet.* This is a biofuel that is used for aircrafts and has properties that are comparable to fossil-based jet fuel. The feedstock used to produce SAF includes municipal solid waste, cellulosic waste, various other energy crops and green hydrogen (Sustainable Aviation Fuel 2020). Currently used process pathways for SAF synthesis includes Fischer–Tropsch Synthetic Paraffinic Kerosene (FT-SPK, using forestry residues and municipal solid waste), Hydro-processed Fermented Sugars to Synthetic Iso-paraffins (HFS-SIP, using microbial conversion of sugars to hydrocarbons), Alcohol-to-Jet Synthetic Paraffinic Kerosene (ATJ-SPK, using agricultural waste and crop straws), FT-SPK with aromatics (FT-SPK/A, municipal solid waste, agricultural wastes and forestry residues, wood and energy crops), Hydro-processed Esters and Fatty Acids Plus (HEFA+, biomass with high oil content including algae, camelina, carinata), Hydro-processed Esters and Fatty Acids (HEFA-SPK) (Beginners guide to Sustainable Aviation Fuel 2017). Obviously due to the varying feedstock and related land use, logistics and processing technologies, the carbon impact is different for each. Also, the economic feasibility and carbon reduction potential is dependent on the region the SAF fuel ecosystem and value chain is situated in. The use of SAF is increasing rapidly globally, and in 2021, it has been evaluated that 70 million gallons of SAF was produced in 2021, and the forecast is 2.6 billion gallons by 2025 (Grimme 2022). The total global jet fuel market is current about 110 billion gallons and is estimated to grow to 230 billion gallons by 2050.

Renewable gasoline is a *green transportation fuel* that uses biomass as feedstock and is chemically the same as petroleum-derived gasoline. It can be used in any conventional gasoline using vehicle and also the gasoline transportation infrastructure works as is. There is only a limited production capacity for this type of fuel but most large oil companies are now looking at transforming their production processes to use bio-based feedstocks to also produce renewable gasoline, renewable diesel, and SAF. Due to for example the COVID-19 pandemic the overall biofuels R&D public funding started declining. The US EPA, USDA and DOE have, however, recently announced increased funding opportunities to better support the developments of the second-generation biofuels' development and their deployment in order to reach the carbon neutrality targets. Emphases are put to demonstrate impactful greenhouse gas savings. The EU is also implementing directives on Biofuels and Fuel Quality that are targeted at preventing incentives for food crop use for fuel (EU Fuel Quality Directive 98/70/EC).

In this chapter, we discussed technology platforms that are in general agreement to be some of the most impactful for mitigating climate change. The latest IPCC report (AR6 WGII) reviews future climate change adaptation options and their feasibility by looking at the systems transitions including land and oceans ecosystems (e.g., coastal defense and zone management, land, forestry, aquaculture, agroforestry, biodiversity, water use management, cropland management and livestock systems), urban and infrastructure investments (e.g., green infrastructure, urban planning, and water management), energy systems (e.g., resilient and reliable energy systems, improved water use efficiency), and cross sectorial issues (e.g., relocation, human resettlement, risk and catastrophe management). This kind of systems view looks at the climate change impacts and risks short and long term and identifies the areas that need to be addressed in order to adapt. In the next chapter we will also look at the available data and will review investments in some of these main technology platforms, including R&D investments, mergers and acquisitions and patenting activity.

References

Advisory Committee for Environmental Research and Education (2018) Sustainable Urban Systems: Articulating a Long-Term Convergence Research Agenda. A Report from the NSF Advisory Committee for Environmental Research and Education. Prepared by the Sustainable Urban Systems Subcommittee

Agora Energiewende (2021) No-regret Hydrogen: charting early steps for H_2 infrastructure in Europe. Available at: https://static.agora-energiewende.de/fileadmin2/Projekte/2021/2021_02_EU_H2Grid/A-EW_203_No-regret-hydrogen_WEB.pdf

Asif M, Suleman M, Haq I, Jamal SA (2018) Post-combustion CO_2 capture with chemical absorption and hybrid system: current status and challenges. Greenhouse Gas Sci Technol 8:998–1031. https://doi.org/10.1002/ghg.1823

Balat M (2011) Production of bioethanol from lignocellulosic materials via the biochemical pathway: a review. Energy Convers Manage 52(2):858–875

Beginners guide to sustainable aviation fuel (2017) Air Transport Action Group. https://aviationbenefits.org/media/166152/beginners-guide-to-saf_web.pdf

Borchers M, Keller K, Lott P, Deutschmann O (2021) Selective catalytic reduction of NOx with H_2 for cleaning exhausts of hydrogen engines: impact of H_2O, O_2, and NO/H_2 ratio. Ind Eng Chem Res 60(18):6613–6626. https://doi.org/10.1021/acs.iecr.0c05630

Bronner W, Gebauer H, Lamprecht C, Wortmann F (2021) Sustainable AIoT: how artificial intelligence and the Internet of Things affect profit, people, and planet. In: Gassmann O, Ferrandina F (eds) Connected business. Springer, Cham. https://doi.org/10.1007/978-3-030-76897-3_8

Carbon Black Market (2021). https://www.fortunebusinessinsights.com/industry-reports/carbon-black-market-101718

Collodi G, Azzaro G, Ferrari N, Santos S (2017) Techno-economic evaluation of deploying CCS in SMR based merchant H_2 production with NG as feedstock and fuel. Energy Procedia 114(2017):2690–2712. https://doi.org/10.1016/j.egypro.2017.03.1533

Copeland BJ (2023) Artificial Intelligence. Encyclopedia Britannica, 16 Feb. 2023. https://www.britannica.com/technology/artificial-intelligence. Accessed 20 Feb 2023

EPA (2022) U.S. Environmental protection agency, inventory of U.S. Greenhouse gas emissions and sinks: 1990–2020, April 2022. Includes U.S. Territories

EWEA (2009) The economics of wind energy. EWEA, Brussels

Ganguly A, Brown RC, Wright MM (2022) Techno-economic and greenhouse gas emission assessment of carbon negative pyrolysis technology. Green Chem 24:9290–9302
Global CCS Institute (2023). https://co2re.co/FacilityData
Global Hydrogen Review (2022) IEA Publications. https://iea.blob.core.windows.net/assets/c5bc75b1-9e4d-460d-9056-6e8e626a11c4/GlobalHydrogenReview2022.pdf
Grimme W (2022) The introduction of sustainable aviation fuels-challenges and options. In: Conference proceeding–IAMC 2022, pp 181–190
Hydrogen roadmap Europe (2019) Fuel cells and hydrogen 2 joint undertaking, hydrogen roadmap Europe: a sustainable pathway for the European energy transition, Publications Office, 2019. https://doi.org/10.2843/341510
Hydrogen-Powered Aviation (2020) A fact-based study of hydrogen technology, economics, and climate impact by 2050. https://www.euractiv.com/wp-content/uploads/sites/2/2020/06/20200507_Hydrogen-Powered-Aviation-report_FINAL-web-ID-8706035.pdf
IEA (2019) Nuclear power in a clean energy system. IEA, Paris. https://www.iea.org/reports/nuclear-power-in-a-clean-energy-system
IEA (2021) Net zero by 2050, IEA, Paris. https://www.iea.org/reports/net-zero-by-2050
IEA (2022) Wind Electricity, IEA, Paris. https://www.iea.org/reports/wind-electricity
Jones AC, Lawson AJ (2022) Carbon capture and sequestration (CCS) in the United States. Congressional Res Serv R44902. https://sgp.fas.org/crs/misc/R44902.pdf
IMO GHG Strategy (2018). https://unfccc.int/sites/default/files/resource/250_IMO%20submission_Talanoa%20Dialogue_April%202018.pdf
Kazemifar F (2022) A review of technologies for carbon capture, sequestration, and utilization: cost, capacity, and technology readiness. Greenhouse Gas Sci Technol 12:200–230. https://doi.org/10.1002/ghg.2131
Løkke S, Aramendia E, Malskær J (2021) A review of public opinion on liquid biofuels in the EU: current knowledge and future challenges. Biomass Bioenerg 150:106094
Madeddu S et al (2020) Environmental research letters, vol 15 p 124004
Munirathinam S (2020) Chapter six—industry 4.0: industrial Internet of Things (IIOT). Adv Comput 117(1):129–164. ISSN 0065-2458, ISBN 9780128187562, https://doi.org/10.1016/bs.adcom.2019.10.010
Nadel S (2019) Electrification in the transportation, buildings, and industrial sectors: a review of opportunities, barriers, and policies. Current Sustain Renew Energy Rep 6. https://doi.org/10.1007/s40518-019-00138-z
Our World in Data (2021). https://ourworldindata.org/co2-and-other-greenhouse-gas-emissions
Peterson OG, Kimpland RH, Coates DM (2008) Compact, self-regulating nuclear reactor. Trans Am Nucl Soc 98:729–730
Porcheron F, Gibert A, Mougin P, Wender A (2011) High throughput screening of CO_2 solubility in aqueous
Rahardjo BS (2013) The assessment of syngas utilization by Fischer Tropsch synthesis in the slurry—bed reactor using Co/SiO_2 catalyst. Int J Eng 4(1):8269
Sausen R, Schumann U (2000) Estimates of the climate response to aircraft CO_2 and NOx emissions scenarios. Climatic Change 44:27–58. https://doi.org/10.1023/A:1005579306109
Shepard JU, Pratson LF (2022) The myth of US energy independence. Nat Energy 7:462–464. https://doi.org/10.1038/s41560-022-01053-2
Sintef (2019) Report: hydrogen for Europe. https://www.sintef.no/globalassets/sintef-energi/hydrogen-for-europe/hydrogen-for-europe-pre-study-report-version-4_med-omslag-2019-08-23.pdf
Sustainable Aviation Fuel—Review of Technical Pathways. DOE, Office of Energy Efficiency and Renewable Energy. https://www.energy.gov/eere/bioenergy/articles/sustainable-aviation-fuel-review-technical-pathways-report
Tulkki V, Pursiheimo E, Lindroos TJ (2017) District heat with small modular reactors (SMR)
UNCTAD (2017) Review of maritime transport 2017. United Nations publication. Sales No. E.17.II.D.5 New York and Geneva. http://unctad.org/en/PublicationsLibrary/rmt2017_en.pdf

Xu Y, Ahokangas P, Louis J-N, Pongrácz E (2019) Electricity market empowered by artificial intelligence: a platform approach. Energies 12(21):4128. https://doi.org/10.3390/en12214128

Yilmaz C, Wendelstorf J, Turek T (2017) Modeling and simulation of hydrogen injection into a blast furnace to reduce CO_2 emissions. J Clean Prod Vol 154:488–501. ISSN 0959-6526. https://doi.org/10.1016/j.jclepro.2017.03.162.

Zhang A, Tay HL, Alvi MF, Wang JX, Gong Y (2022) Carbon neutrality drivers and implications for firm performance and supply chain management. Bus Strategy Environ 1–15. https://doi.org/10.1002/bse.3230

Chapter 6
Investments in Carbon Neutrality–"Follow-The-Money"

Abstract In this section, we explore the vital role of sustainable banking and finance in supporting the transition to carbon neutrality. We explore the various financial aspects, including government and corporate R&D, infrastructure development, renewable energy investments, clean-tech deployment, private equity, green bonds, and mergers and acquisitions, all of which are instrumental in driving climate policies and targets. With a macroeconomic and global perspective, insights are provided into selected public and private investments, highlighting the need for collaboration between governments, businesses, and investors to mobilize capital towards a greener future. The ambitious net zero emissions 2050 targets are discussed in the context of the urgency of major increases in climate technology investments. To achieve the ambitious targets, there must be a focus on technologies that facilitate the phasing out of fossil fuels. We discuss the investment actions falling behind in all technology platforms, underscoring the pressing need for accelerated action and investment in sustainable solutions.

Keywords Sustainable banking and finance · ESG—environment, social and governance · Public investments · Private investments · Global green bonds · Green energy investments · Patent applications · Emerging technologies

6.1 Background

It is clear that all the global policy changes related to carbon neutrality demand significant investments: corporate and societal. Therefore, financial aspects by all actors are some of the most important issues impacting the implementation of the policies and targets. These *climate investments* include such topics as *government and corporate R&D, city and municipality infrastructure, renewable energy capacity investments, clean-tech deployments, private equity and venture capital, mergers and acquisitions and green bonds.* Finance is critical also in the global *climate change adaptation* schemes also reported in the recent IPCC AR6 report. The report concludes that the positive trajectory of both private and public climate investments

M. Hakovirta, *Carbon Neutrality*, Springer Climate,
https://doi.org/10.1007/978-3-031-45202-4_6

is far behind the foreseen needs for implementation. For example, according to IEA, the investments required in this decade, in order to be on track with the *net zero emissions* global target of 2050, are USD 3,200 billion. However, in 2022 these reported investments amounted only to about USD 1,400 billion.

The missing gap is extremely high at 1,800 billion. However, these investments seem to be slowly picking up and for example global green energy investments have been rising at an estimated 4% annual growth rate (IEA WEO 2022). For better understanding of the situation and the net zero investment portfolios globally, we will discuss sustainable banking and finance and review some high-level insights into the progress of investments by corporations and governments that are driving global decarbonization.

6.2 Sustainable Banking and Finance

The rise of *sustainable banking and finance* is at the core of integrating sustainability into the corporate strategies and business models. It also connects to funding new businesses and raising capital for business growth. Clearly *corporate finance* is becoming increasingly important in further driving and implementing global sustainability and carbon neutrality. The significance of the corporate finance in driving change towards carbon neutrality comes from the fact that most all business decisions with financial implications can be considered as corporate financial decisions.

The business function of corporate finance is related to raising capital and managing the required finances and sources including financing, capital structuring, and investment decisions in the best possible way ethically and economically. In the world of finance, all is connected globally and therefore the relation to the *global financial system* is important to be familiar with. The global financial system works as a large network of different financial institutions for efficient transfer and exchange of capital. And by essentially using this networked system various types of investors receive capital to fund investments in *mergers and acquisitions, organic growth of companies, capital investments, R&D projects, and start-ups.* By these financial resources, they are able to ultimately receive return for their investments and grow the economy. The financial system includes multitude of different actors and stakeholders and consists of, for example, insurance companies, investment banks, commercial banks, government, and mutual funds. The perspective of all these players depends on what level of the system we are looking at. The level or scope includes for example company, regional, national, and global levels.

When we look more closely at *sustainable investing,* it has a number of investing strategies in different shapes and forms. One approach is *socially responsible investing (SRI),* which has become more popular recently and seeks companies that are inclined to look at positive social impact due to the business they are in (e.g., business model, or technology) and how the leadership and governance works. The companies in this space can receive *socially responsible investments* for example through socially conscious mutual funds or exchange-traded funds (ETFs). Although return on investments is still the key aspect in investment decisions in SRI, some of the companies that are funded have strong ties to community investing with emphasis on community impact. Closer look at the SRI investments shows that SRI tends to "follow the times," meaning that it is following timely social, political, and economic trends and bases decisions on those. Since climate change is extremely timely SRI has also become part of the climate-related finance trend.

As mentioned earlier the use of *ESG or environmental, social, and corporate governance standards* are already now essential for any large and institutional investors. The financial investment and capital allocations are directed more towards companies that understand and follow *ESG criteria* with their sustainable business models and long-term vision. It is worth to note that unfortunately the ESG is many times mixed with other sustainable finance concepts such as *SRI, sustainable investing, responsible investing, and impact investing*, all factually having their distinct differences. The difference with ESG is that its criteria are specifically looking at environment, social, and governance topics. The ESG criteria are a bit subjective still, but it is continuing to develop towards more objective and measurable rating. In the past, the corporations with good *investment relations* business functions that were able to tell a *high-impact story* on companies' ESG policies and actions were rewarded. A lot of consulting service companies have benefitted from this trend and are providing services to help corporations create their sustainability and ESG strategies. Examples of *ESG—Environmental criteria* could include how a company uses renewable energy sources in its operations. Or how industrial waste is managed or valorized. Other aspects may involve air or water pollution management, company's supply chain management, and how a corporation looks at biodiversity, conservation, preservation, and deforestation issues. And of course, corporate agendas and business models related to climate change mitigation and decarbonization are increasingly becoming ever more important.

The *social criteria in ESG* look specifically at the key aspects of social relationships between company's employees and stakeholders and can include fair wages, employee benefits, and retirement plans (Schoenmaker and Schramade 2019). It also includes *policies on diversity, equity, and inclusion (DEI) and sexual harassment*. This criterion can also include employee's *education and career advancement* and *related financial support*. Also, *human rights issues, child labor policies,* and *local community involvement and support* are key parts of ESG social criteria.

The third *ESG criteria* governance looks at the overall *corporate governance model*. This includes the *board of directors* and *executive management* and various conducts with stakeholders. An important part of this criteria is the *financial reporting* and *financial and accounting transparency. The fiduciary duty* of company executives and directors is to prioritize the best interests of the shareholders and *includes* for example potential *conflict of interest* aspect. Also, policies on *executive compensation* and *incentive systems* are important for example in looking at how vested the executives are in terms of creating long term and a sustainable competitive advantage. Also, *employee involvement in governance, shareholder,* and *customer satisfaction* are *included* in this criterion. The ESG investing is continuing to develop, and more accountability and responsibility are being included. These advancements do not

Fig. 6.1 Examples of ESG—environment, social, and governance investment criteria

come without challenges. For example, some controversial changes have occurred lately in October 2020. The US Department of Labor then released a new regulation that was limiting socially responsible investing in retirement plans. It undermined certain financially profitable investments. These events for example mandated the fiduciaries of retirement plans to focus more on investment strategies based on only financial performance. The regulations have since then been revisited (Fig. 6.1).

The sustainable finance continues to escalate in its importance in implementing sustainability in corporate settings and in driving carbon neutrality goals. This trend supports the development of more sustainable business models and creates better access to capital in terms of critical technology platform developments for mitigating climate change.

Examples of additional opportunities in capital loans and direct financial support that connect to carbon neutrality developments include support by the *European Investment Bank (EIB)* and *World Bank Group (WBG)*. Both are *institutional financial entities* that support global needs for various issues with a special focus to transition to low-carbon products, and climate change mitigation-related businesses. For example, during 2011–2015 EIB approved EUR 90.5 billion in *climate action lending* and they are *committing to use 25% of all EIB lending for climate-related investments.* This amounts to approximately EUR 100 billion for climate-related projects. The recent commitments are even more ambitious. EIB has an objective of supporting €1 trillion in green financing this decade, and they reported solid progress having already backed €222 billion in investment since 2021 (EIB 2022). The WBG is another major financial contributor to sustainability globally, and it is considered as the prominent funder for developing countries' climate change mitigation efforts. WBG is a *Multilateral Development Bank (MDB)*, which consist of donor and borrowing member countries and offers financing and advice services to advance developing countries. WBG supports industrial ecosystems and markets at a local level. With its latest climate finance strategy, WBG is using a minimum of 28% of its total budget to climate finance. This currently exceeds $20 billion annually.

The WBG's Climate Change Action Plan (2021–2025) is aiming at climate change adaptation, support for developing countries fiscal and financial management and cost-sharing, and leveraging private sector investments to critical and impactful decarbonization and climate change adaptation projects (WBG 2023). Examples are numerous and include a program for boosting the investments to create solutions for wind, solar, and battery storage for reliable continuous availability for electrical power. One of the more particular goals for this program is to help financially to develop 17.5 GWh of battery storage to developing countries by 2025 (currently installed capacity is approximately 5 GWh). Other example is the support by WBG to India's goal to increase its renewable energy capacity from about 40 GW to 175 GW by 2022. The focus is especially on solar power. India has an ambitious goal to increase its solar power by 100 GW. This project is however behind schedule since as of June 2022 the total solar power was 57.7 GW. At the same time, the total electricity production installed capacity was 403.8 GW (WBG 2023). *International Finance Corporation (IFC) is also* part of WBG and is the largest global development institution. Its aim is to support and advance the private sector in the developing countries. The total trade-finance volume of IFC is $211 billion. They operate also in climate finance and have so far issued green bonds that have raised more than $20 billion to support projects that are advancing decarbonization (IFC 2022).

Another large climate finance leader is the *Climate Investment Funds (CIF)* that is composed of *Clean Technology Fund (CTF)* and *Strategic Climate Fund (SCF).* Established in 2008, in total CIF has over $10 billion in funds used together with various partners. The approach is to use the funds by leveraging and managing various programs for climate change mitigation (the total scope is 325 projects in 72 countries globally). The total climate finance impact has been about $60 billion and includes partnerships with WBG, and other MDBs together with private sector support. These efforts have helped access to renewable energy, develop sustainable forestry practices, and prepared developing countries for climate change adaptation.

6.3 Public and Private Investments Insights

Decarbonization-related investments can be categorized to be from governments, financial institutions, or corporations. The governmental and financial institutions include topical examples such as public R&D spending, city and municipality infrastructure, and green bonds. Included are also capacity investments and clean-tech deployments. Corporate investments encompass a range of activities, including private sector R&D, mergers and acquisitions, private equity, and venture capital. Also, patent applications in terms of climate technologies are included to these numbers. A closer look at these categories gives a so-called sectoral and technology platform-related dimensions. As stated earlier, the average annual investments required to reach net zero 2050 targets in terms of *green energy investments are far from being on track.* Advanced economies are clearly driving green energy investments on a global scale and emerging and developing economies are far behind even when the COVID-19 pandemic and increased energy prices are factored into the equation. Looking at the largest contributors specifically to the *energy transition spending*, the dominant nation is China (35%), followed by the European Union

(20%), and the USA (15%). The spending growth is clearly highest in Europe at 48% (compound annual growth rate or CAGR, 2019–2021). The second highest growth is in China at 19% CAGR (2019–2021), and the lowest growth during the same period was 2% CAGR in the USA (Fig. 6.2).

According to the *Climate Policy Initiative*, climate-related investments are strikingly equally distributed by both public and private actors. These investments aim at reducing emissions and improve adaptation and resilience to climate change. It is, however, estimated that moving forward this balance will shift quite a lot and 2/3 of the necessary investments for net zero emissions 2050 target will eventually come from the private sector (Climate Policy Initiative 2022) (Fig. 6.3).

Climate change mitigating technologies are considered to reduce and avoid GHG emissions or to be able to enhance the GHG sequestrations. On the other hand, *adaptation* refers to endeavors aimed at decreasing the vulnerability of human or

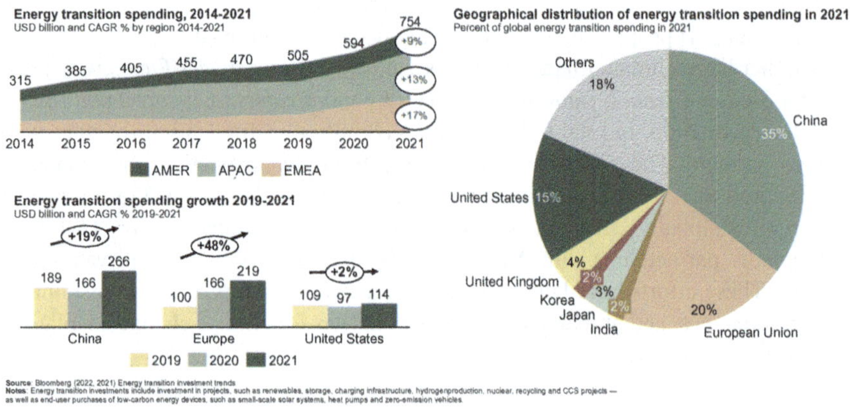

Fig. 6.2 Global contributors to energy transitioning spending

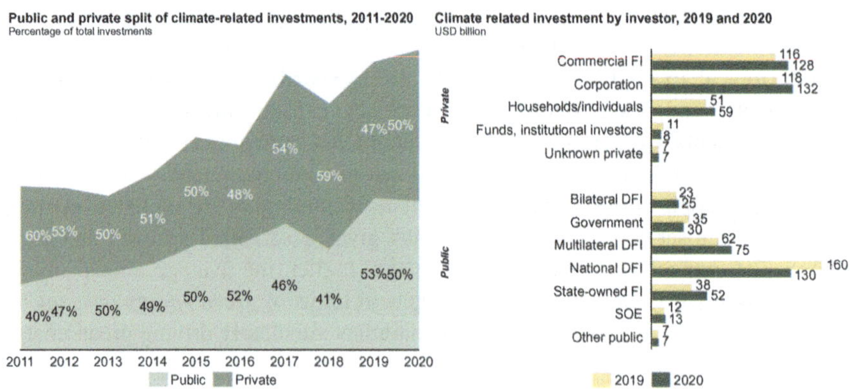

Fig. 6.3 Climate-related investments by actors (Climate Bonds Initiative 2022)

natural systems to the impacts of climate change and associated risks. The share of climate-related investments used for climate change mitigation or adaptation in 2021 shows that the percentage of total climate-related investments related to mitigation was 89% and adaptation only 8% (Climate Policy Initiative 2022). However, the investments to adaptation will increase significantly as the climate change impact increases. World Economic Forum estimates this to be a $2 trillion market per year by 2026, with the increased growth as climate impacts become more prevalent (WEF 2022).

From industrial sector perspective, the five largest climate-related investments in 2021 were energy (53%, USD 354 billion), transportation (24%, USD 162 billion), and buildings and infrastructure (9%, 58 billion). The total investments in 2021 were USD 665 billion. Although the transportation electrification-related investments are growing rapidly (59% CAGR during 2019–2021), the renewable energy continues to be the highest. Its growth was, however, only 11% during 2019–2021 (IEA WEI 2022; Bloomberg 2022).

Government investment and *green bonds* are also a critical part of green transformation towards carbon zero targets. Especially education and research sectors are highly reliant on the governmental support. Unfortunately, according to IEA (2022), governmental spending on energy technology R&D has been slightly declining in many advanced economies globally, with the exception of the EU, UK, and USA. Especially EU is increasingly adding to its public R&D funding for energy technology (8% CAGR increase, 2015–2020) (Fig. 6.4).

The situation is slightly more promising looking at global level as governmental energy technology R&D spending has continued to increase at a modest pace since 2015. The most spending has been on energy efficiency projects, followed by nuclear fission and fusion and renewable energy-related R&D investments (IEA WEI 2022) (Fig. 6.5).

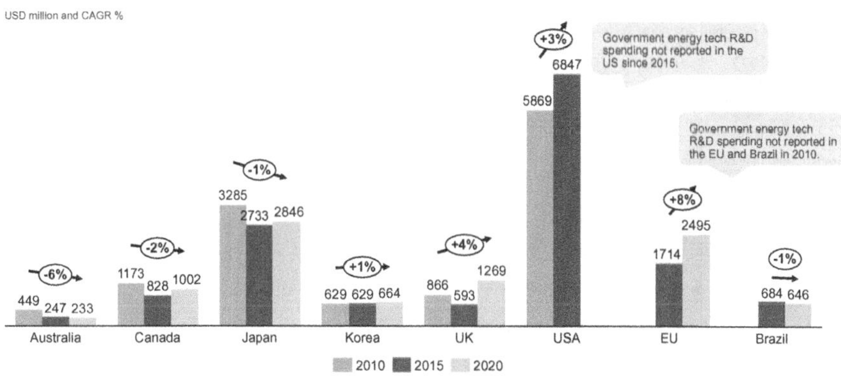

Fig. 6.4 Governmental spending on energy technology research & development (R&D) by country, 2010–2020 (IEA GET 2022)

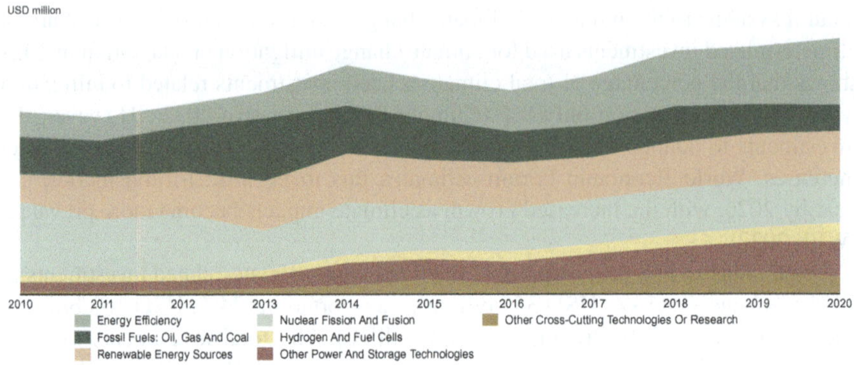

Fig. 6.5 Governmental spending on energy technology research & development (R&D) by technology, 2010–2020 (IEA GET 2022)

Some of the notable trends in these investments can be seen in key sectors and technologies, such as *energy efficiency* (e.g., transportation, 24% CAGR increase) and *power and storage technologies* (e.g., electric power generation, 26% CAGR increase, and power-energy storage, 12% CAGR increase) have seen a high increase in governmental energy technology R&D spending since 2015. Not surprisingly, the fossil fuel technologies have been in decline, especially coal, oil, and gas. This is unfortunate as we still remain reliant on these sectors for decades and increasing energy efficiencies and emissions remain vital for mitigating GHG emissions. The impact of the technological improvements in these energy sectors still remains high and in the short term may even be higher than the long-term alternatives development. Another topic of concern is the decline in CO_2 capture and storage public R&D investments. In this case, there is a decline of 4% CAGR during 2015–2020 (Fig. 6.6).

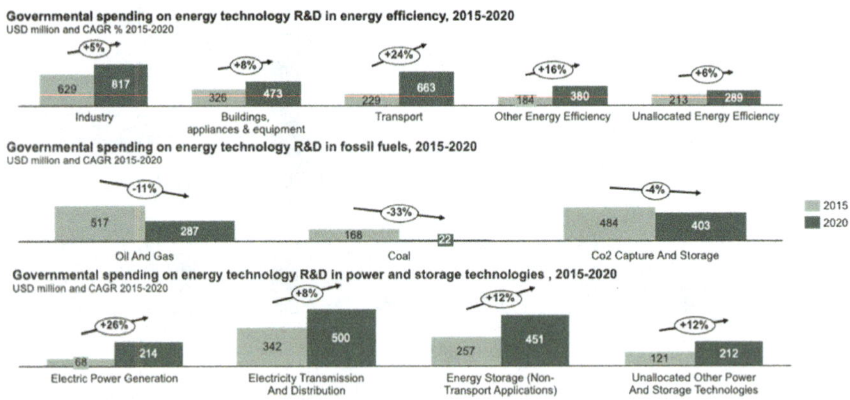

Fig. 6.6 Governmental spending on energy technology R&D in energy efficiency, fossil fuels, power, and storage technologies, 2015–2020 (IEA GET 2022)

One of the major trends in sustainable finance has been the *green bonds development*. These fixed-income financial instruments are issued with the specific purpose of raising funds for projects that have a significant positive impact on climate change and the environment. From a technical standpoint, these bonds are primarily asset-linked, meaning they are supported by the balance sheet of the issuing entity. This structure provides the advantage of carrying the same credit rating as any other debt obligations issued by the same entity. Depending on the issuer or the jurisdiction of the issuing location, the green bonds can also offer tax exemption or tax credits as an incentive. These bonds were originally issued by large financial institutions and entities such as WBG and EIB; however, increasingly local, and municipal governments and non-financial corporates are entering the market.

In recent years, there has been a significant increase in the global issuance of green bonds. The trend shows financial and non-financial corporations issuing these bonds to projects in the energy and buildings sectors. Financial corporations account for 26%, non-financial corporations, 26%, sovereign bonds 15% (bond issued by a government to support public spending), government-backed entities 15%, development banks 7%, and asset-backed securities (bonds backed by a portfolio of debt instruments) 5% (Climate Bonds Initiative 2022) (Fig. 6.7).

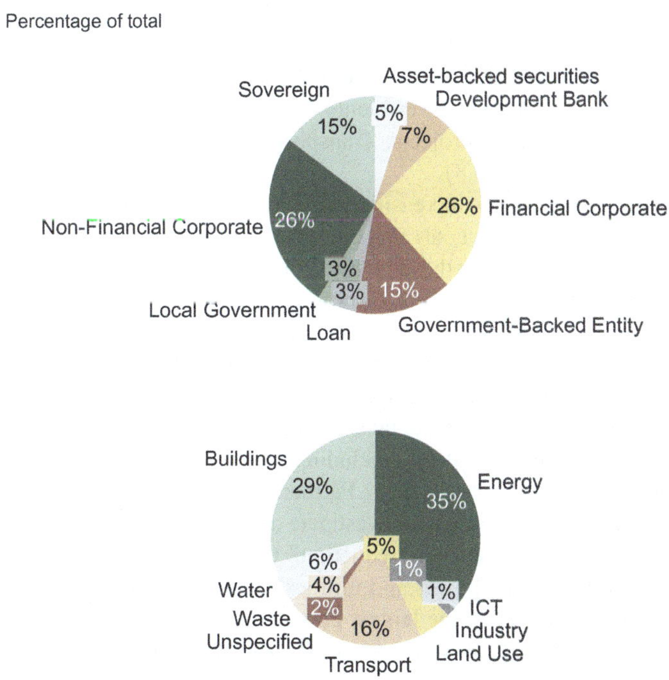

Fig. 6.7 Global green bond issuance by issuer type, and use by sector (Climate Bonds Initiative 2022)

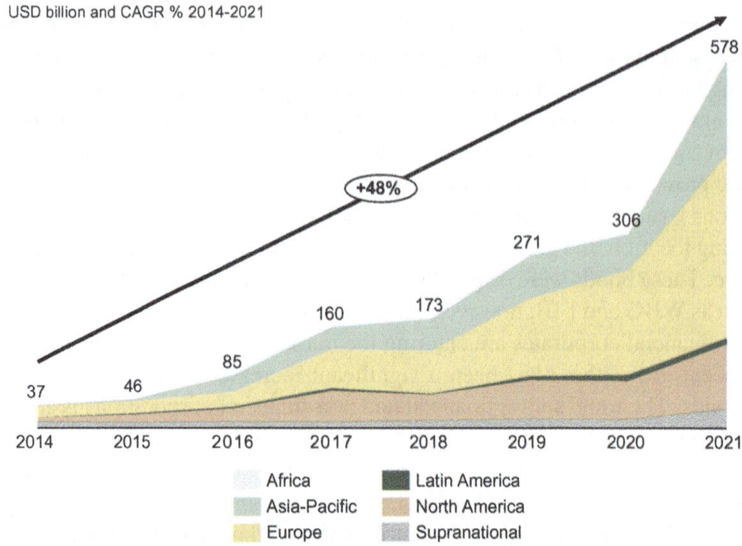

Fig. 6.8 Value of global green bonds issued by region, 2014–2021 (Climate Bonds Initiative 2022)

The total amount of green bonds was in 2021 at USD 578 billion. Europe is in the leading position in green bonds development, followed by North America and Asia–Pacific. The green bond investments are clearly dominated by energy (35%) and building (29%)-related projects. Transportation is at 16%, followed by water (6%) and waste (4%) (Fig. 6.8).

Corporate investments are increasingly important in order to reach the demanding net zero or carbon neutrality goals that corporations have been committed to during the past few years. Drivers for these investments are evidently not just about participating into global movement towards carbon zero by 2050. However, there are clearly tremendous business opportunities that the ambitious net zero goals are laying out and additionally corporations see significant business risks associated with climate change that demand necessary investments to climate adaptation.

GlobalData disruptor platform database is a databank that includes information on global private sector financial deals (including *Venture Capital (VC), Private Equity (PE), Mergers and Acquisitions (M&A)*, Partnerships, debt offerings, and asset transactions) with analytics tools to also investigate different themes (GlobalData 2023). If we examine "carbon emissions reduction" theme in the GlobalData database, corporate deals under the "power and utilities" sector are clearly dominant (USD 125 billion, 2022). However, the largest growth by far comes from the automotive sector. Since 2018, these deals have grown 525% CAGR. The next largest growth is in mining (87% CAGR), technology, media, and telecommunications sector (72% CAGR), and financial services (60%). Interestingly industrial goods and machinery have grown only 35% CAGR (GlobalData 2022) (Fig. 6.9).

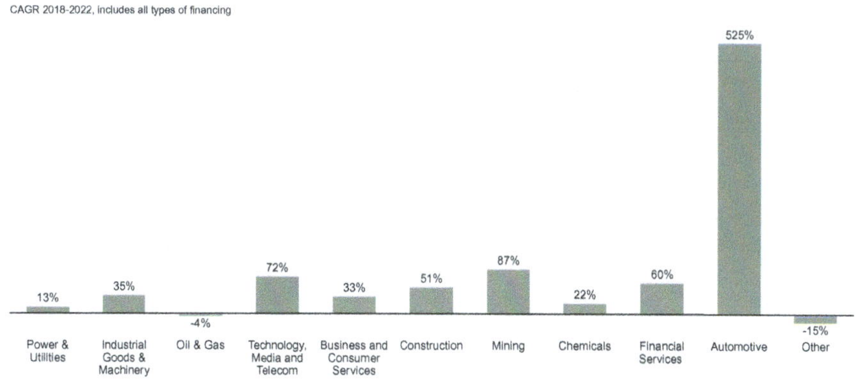

Fig. 6.9 Growth in total value of carbon emissions reduction-themed deals by sector, 2018–2022 (GlobalData 2022)

In terms of geographical distribution, most "carbon emissions reduction"-themed deals have occurred in Europe, followed by East Asia and North America and (GlobalData 2022). The share of East Asia has steadily increased as the North American investments share of the total has been reduced. According to the data during the year 2021, there was an exceptional downturn with about 1/3 of the 2020 investments. This was most likely due to the economic slowdown caused by COVID-19 pandemic (Fig. 6.10).

Patenting activity is in general considered as one of the key metrics to measure corporation's innovativeness. However, on the one hand it does encourage innovation by giving an inventor an exclusive protection for an invention for some 20 years. On the other hand, the purpose of patenting is to create a barrier for someone else to use the patented technology. This means also that there may not be an intention

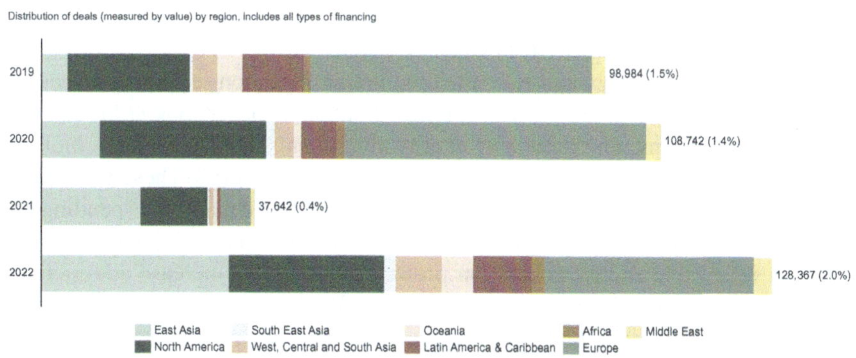

Fig. 6.10 Distribution of total carbon emission reduction-themed deals (measured by value) by region, including all types of financing, 2019–2022 (GlobalData 2022)

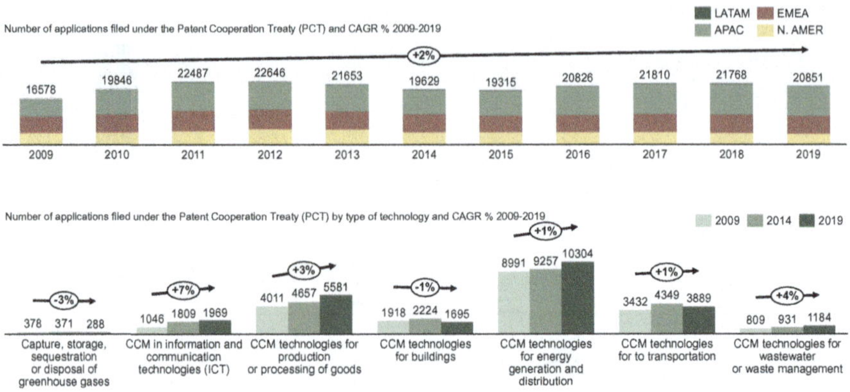

Fig. 6.11 Patent applications filed for climate change mitigation (CCM) technologies globally, 2009–2019 (OECD (2022) Climate Change Mitigation patents database)

to develop, commercialize, or defend the patented technology. Regardless of this dilemma, observing patenting data, one can see trends in various technology platform developments and the research intensity.

According to OECD Climate Change Mitigation patent database, innovation in climate change technologies has been growing relatively constant (2% CAGR) over the last decade, with the largest growth observed in ICT sector (OECD 2022) (Fig. 6.11).

Upon closer examination of the data, it becomes evident that the highest level of patenting activity has been focused on CCM technologies related to energy generation and distribution (10,304 applications in 2019). The second-largest activity has been in technologies for production or processing of goods (5581 applications in 2019), and the third largest has been for transportation technologies (3889 applications in 2019). It is important to note that all these technology areas have grown only slightly (1–3% CAGR, 2009–2019). The largest growth has been in information and communication technologies (ICT) with 7% CAGR.

Since renewable energy and electrification in transportation sectors have seen the highest levels of investment growth, it is worth taking a closer look at the relevant technology platforms. The renewable energy investments include biofuels, biomass and waste to energy, small hydropower, solar, and wind energy. These technology investments have grown at a CAGR of 7% in the last decade, and the spending has mostly been on *capacity expansion*. The term capacity expansion refers to investments in renewable energy generation projects for capacity increase and includes both asset finance and utility-scale projects (UNEP 2020).

More specifically solar and wind energy dominates the investment landscape for renewables—both in terms of absolute levels and growth. In 2019, both were at USD 140 billion level with a growth level of 7–8% CAGR. The other technology areas have faced negative growth (UNEP 2020) (Figs. 6.12, 6.13 and 6.14).

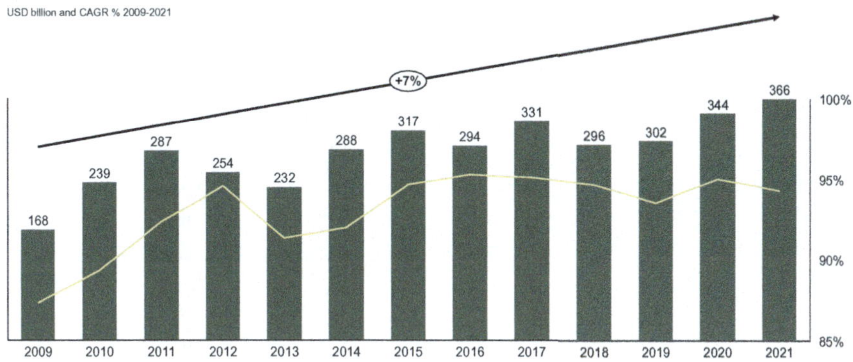

Fig. 6.12 Global renewable energy investment and share of investment that has been used for capacity expansion (right axis and yellow line), 2009–2021 (UNEP 2020)

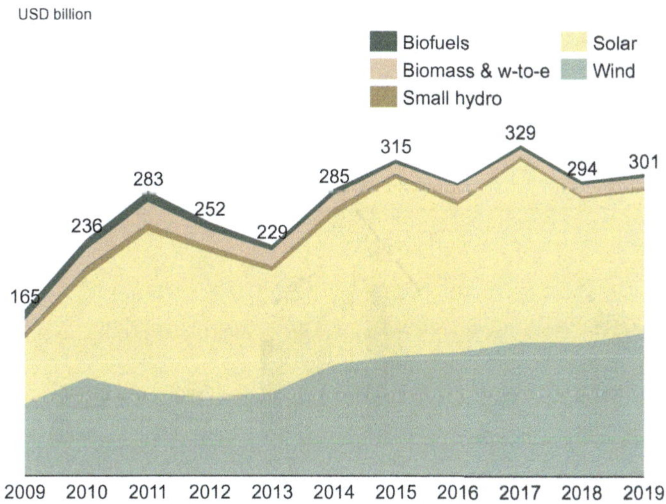

Fig. 6.13 Share of total renewable energy investment spent on each technology, 2009–2019 (UNEP 2020)

In both 2019 and 2020, there was a clear decline in renewables investments in many parts of the world. The only exceptions were USA, Middle East and Africa, and India. This is mostly due to economic slowdown-related lower investment activity due to COVID-19 (UNEP 2020). The renewables investments have picked up speed globally since 2020; however, more detailed data was not available at the country/technology level during the timing of writing this book (Fig. 6.15).

Electrified transportation is expected to overtake renewable energy investments in 2023 (IEA EV 2022). One of the most important indicators for the advancement of transportation electrification is the growth of charging stations. The number of

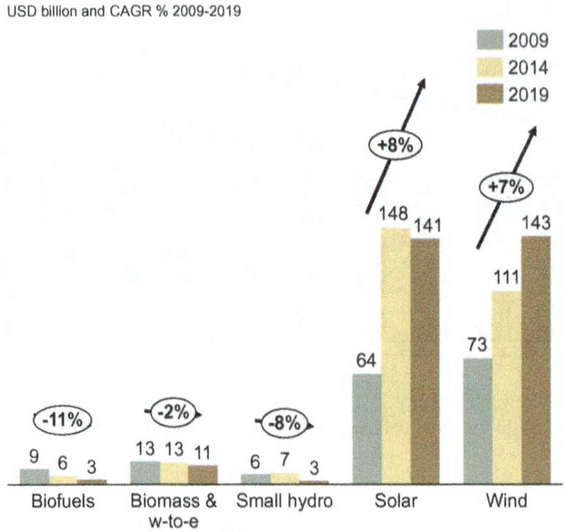

Fig. 6.14 Global renewable energy investment by type of technology, 2009, 2014, 2019 (UNEP 2020)

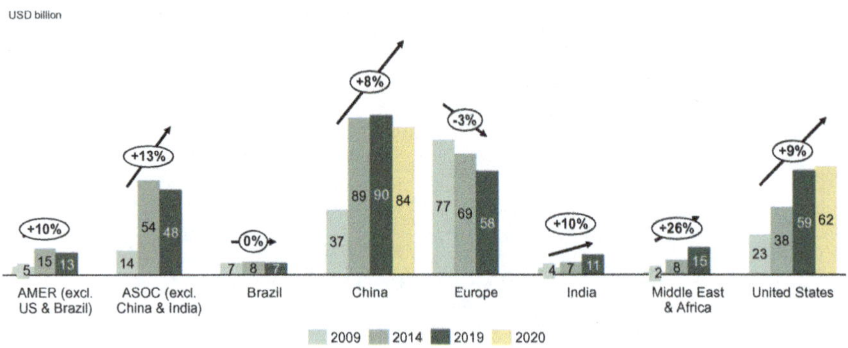

Fig. 6.15 Renewables investments globally and in select countries/regions, 2009–2019 (2020 if available) (UNEP 2020)

public EV charging stations has indeed exploded globally in the last decade, with China accounting for most of the growth and about 60% of all stations since 2019 (IEA 2022). The highest country-specific growth came from Korea (705%), Italy (696%), and Finland (673%) (Fig. 6.16).

In terms of electrification of public transportation, China has by far had the largest stock of electric buses in the world. Europe has been growing its stock by 45% CAGR rate since 2017. Other countries, such as Korea, are ramping up stock at a rapid pace (Fig. 6.17).

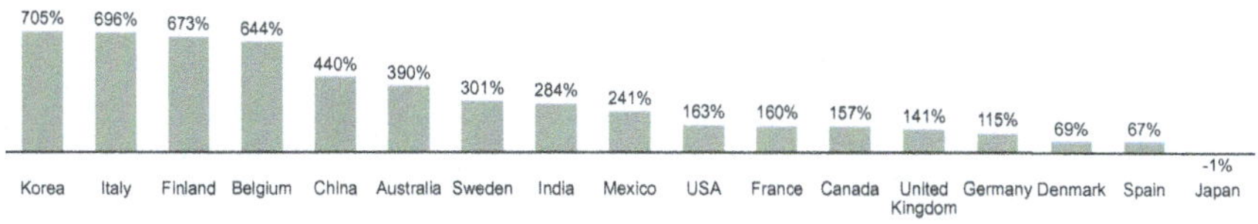

Fig. 6.16 Growth of publicly available EV charging stations between 2017 and 2021 (IEA EV 2022)

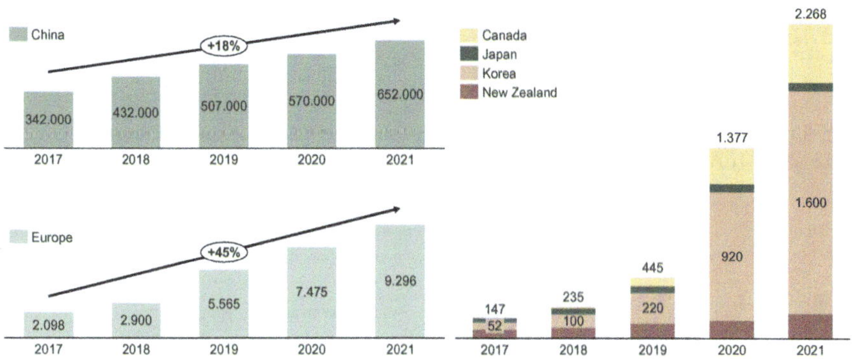

Fig. 6.17 Number and growth percentage (CAGR) of electric bus fleet (battery electric and plug-in) in different parts of the world, 2017–2021 (IEA EV 2022)

Emerging markets and developing economies are also in the progress of ramping up their electric bus fleet. The growth between 2017 and 2020 has been as high as 188% including for example Chile, Columbia, Kazakhstan, Brazil, and Thailand. This is important as their public transportation fleet is extremely outdated and highly polluting (ICCT 2022).

Emerging technologies are needed to reach the 2050 target, and therefore, emerging technology investments are important to follow. Looking more closely at this category, including carbon capture, utilization, and storage (CCUS) and low-carbon fuels (mainly biofuels and hydrogen), the data shows that the investments have quadrupled (41% CAGS) since 2018, but total investment still remains low in absolute terms (IEA 2022). It is likely that both corporate and governmental net zero pledges in combination with strong increase in hydrogen demand may trigger additional investment into CCUS. It is also expected that the USA is taking a larger role leading up to 2030. It is noteworthy to mention that although the total CCUS investments have been increasing, the worldwide government funding for CCUS technology R&D has been declining by 4% CAGR since 2015 from USD 484 billion down to USD 403 billion in 2020 (IEA GET 2022) (Figs. 6.18 and 6.19).

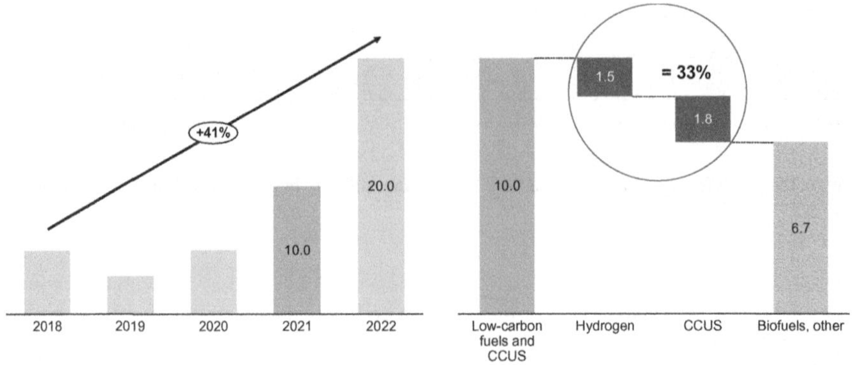

Fig. 6.18 Global low-carbon fuels and CCUS investment (USD billion and percentage CAGR), 2018–2022 (IEA 2022)

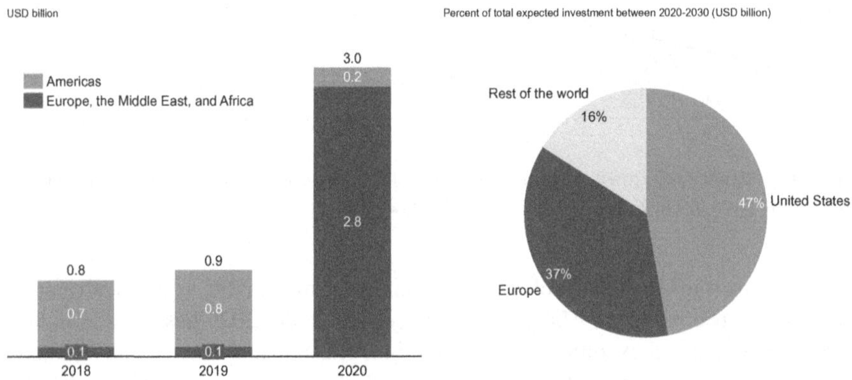

Fig. 6.19 Investments (USD billion and percentage) into CCUS worldwide by region 2018–2020 (IEA 2022)

The hydrogen technology is one of the "wild cards" in the climate technology investments and interestingly the growth of *electrolyzer capacity* is an indicator of hydrogen economy's overall growth expectations. During the recent years, there has been a substantial increase (170%) in electrolyzer installations. This progression is led by China and Europe and is expected to continue towards 2030. A rapid growth is critical for accelerating the uptake of low-carbon hydrogen economy (IEA WEI 2022) (Figs. 6.20 and 6.21).

Although hydrogen has major potential as a low-carbon fuel, however, it still demands several evolutionary steps to become an established and feasible concept and solution. The positive growth in government R&D spending into hydrogen technologies has continued and has been 33% CAGR during 2015–2020 with a global total of $518 billion in 2020. There are also several large-scale government programs globally to support hydrogen demonstration projects (Table 6.1).

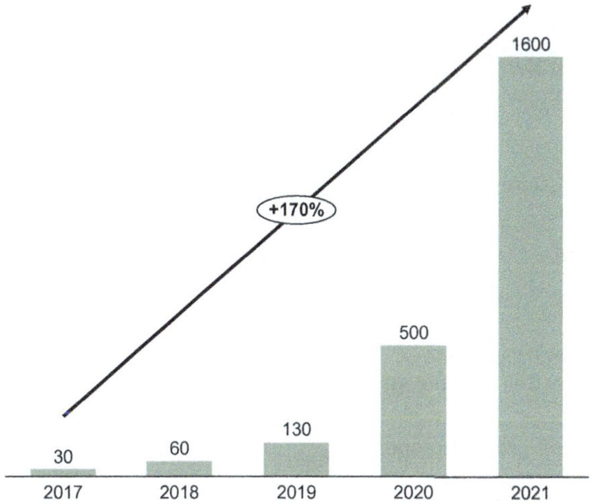

Fig. 6.20 Annual investment in electrolyzer installations, 2017–2021. USD million, CAGR % 2017–2021 (IEA WEI 2022)

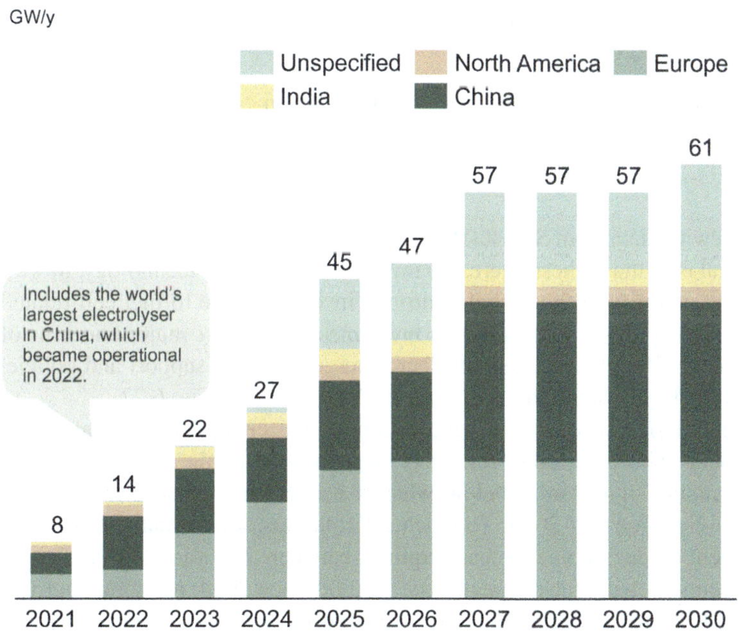

Fig. 6.21 Planned electrolyzer manufacturing capacity by region, 2021–2030 (IEA WEI 2022)

Table 6.1 Government programs to support hydrogen demonstration projects

Programme	Country/region	Value
Green Innovation Fund	Japan	USD 16 billion
Innovation Fund	European Commission	USD 3.2 billion
Bipartisan Infrastructure Law - R&D funding provisions	United States	USD 1 billion
GroenvermogenNL	Netherlands	USD 886 million
Hydrogen Flagship Projects	Germany	USD 828 million
National Innovation Programme Hydrogen and Fuel Cell Technology	Germany	USD 422 million
PERTE renewables, green hydrogen and storage	Spain	USD 296 million
Regulatory Sandboxes for the Energy Transition	Germany	USD 198 million
HyGate	Germany, Austria	USD 89 million
Energy Technology Development and Demonstration Program	Denmark	USD 88 million
Low Carbon Hydrogen supply 2	United Kingdom	USD 75 million
Industrial Fuel Switching	United Kingdom	USD 75 million

Reviewing data from Statista (Statista 2021), it is evident that the transport sector in general is of the highest interest to investors in hydrogen technology. More specifically the majority of the global hydrogen investments go to fuel cell-related technologies. Despite some increases in investments, there are major concerns of costs and profitability for hydrogen and therefore government support and policies have and will be key to facilitate its growth (IEA GET 2022) (Fig. 6.22).

To summarize, while there are indications of a recent uptick in global clean energy investments after a period of relatively stagnant growth, the current level of investment remains significantly below what is needed to stay on track for achieving net zero emissions by 2050. To reach 2050 targets, an estimated USD 32 trillion investment in decarbonization is required between 2020 and 2030. Clean energy investments globally have shown a 4% CAGR since 2014, reaching approximately USD 1400 billion in 2022. While the growth is positive, the current investments fall considerably short of the average annual investment target of USD 3200 billion for this decade. Consequently, there is a clear need for increased global efforts to attract more substantial investments in decarbonization initiatives.

Over the past years, the distribution of climate-related primary investments between private and public actors has remained relatively equal and steady. However, the economic prospects of low-carbon investments, particularly in solar PV and EVs,

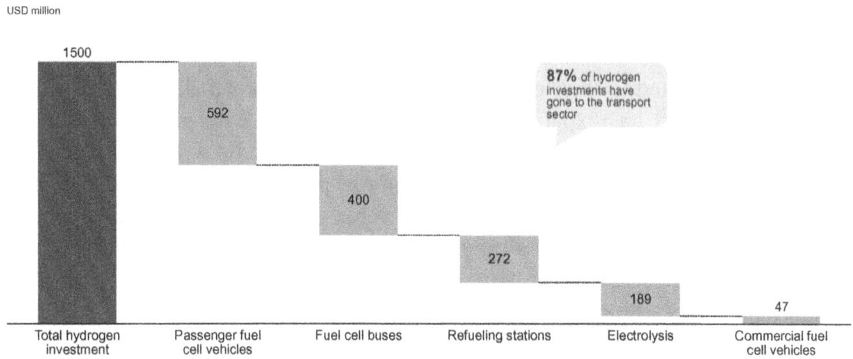

Fig. 6.22 Global Investment (USD million) into hydrogen by type of technology, 2020 (Statista 2021)

are rapidly improving. Further, it is estimated that private actors, especially corporations, could potentially provide up to 70% of the global investments required for achieving net zero emissions by 2050 (40% currently). Nevertheless, it's important to emphasize that public finance and a solid policy framework are crucial in stimulating and facilitating private investment. Public finance plays a pivotal role in incentivizing private investment by offering subsidies, incentives, and financial mechanisms that mitigate risks and enhance returns. Furthermore, a robust policy framework is essential to provide stability and clarity, creating an environment that encourages private investors to engage in sustainable and low-carbon ventures.

Decarbonization investments have been concentrated in advanced economies, while emerging and developing economies have lagged behind. China, in particular, has emerged as a major driver of investment growth in decarbonization, becoming the largest contributor globally. Between 2019 and 2021, China saw a remarkable 19% CAGR in energy transition investments, reaching USD 266 billion in 2021. The IEA predicts that a significant portion of global solar and wind energy growth in the next five years, around 36% and 40%, respectively, will come from China. This highlights China's crucial role in shaping the expansion of renewable energy on a global scale.

Environmental factors are now a central focus in global capital markets, leading to the rapid emergence of new funds and financing instruments. One prominent development is the notable increase in the issuance of green bonds. In 2021, the total value of globally issued green bonds reached USD 578 billion, with Europe accounting for 50% of the market, followed by North America and the Asia–Pacific region. These bonds are issued by various investors, particularly financial and non-financial corporations, to finance decarbonization projects across multiple sectors. Notably, in 2021, 29% of green bond proceeds were allocated to the buildings sector, while the energy sector received 35% of the total proceeds.

Renewable energy has been the dominant focus of low-carbon investments, attracting USD 445 billion in 2021. The declining costs of solar and wind technologies have contributed to their competitiveness and widespread adoption. However, electrified transport has experienced remarkable growth, with global investments reaching USD 273 billion. This sector is projected to catch up with or even surpass renewable energy investments by 2023. Improved efficiency and economies of scale are driving this expansion, and electric vehicles are expected to become cost-competitive with high-carbon-emitting cars by 2026. These developments may indicate a changing landscape where both renewable energy and electrified transport play crucial roles in decarbonization efforts.

From technology platform perspective, the highest investment volumes and still highest growth come from both solar energy and wind energy. The emerging hydrogen technology has great expectations and therefore very high investment growth rates over the last few years. From volume perspective, these still are relatively low investments. Electric vehicles have both very high investment volumes and growth rates. This relates mainly to passenger vehicles, but also to public transport electrification, especially in China. The energy storage technology has vigorous investment growth rates, led by the USA, also the CCUS technology has very high investment growth rates over the past few years. However, the total CCUS investment volumes are still low. Publicly funded fuel cell R&D investments have been stagnant globally since 2015. Nonetheless, looking at transport sector's hydrogen investments, majority is focused on fuel cell vehicle-related technologies. It is also good to note that 87% of hydrogen investments have gone to the transport sector. Electricity grids have high investment volumes; however, the growth seems somewhat dormant worldwide. High investment volumes exist for electrified heating, although this technology platform has slow growth. Biomass and waste-to-energy investments have currently higher investment volumes than biofuels and small hydro but are still lacking growth. Interestingly, biofuels have faced a slightly declining investment growth during the past recent years and also investment volumes are now fairly low compared to the

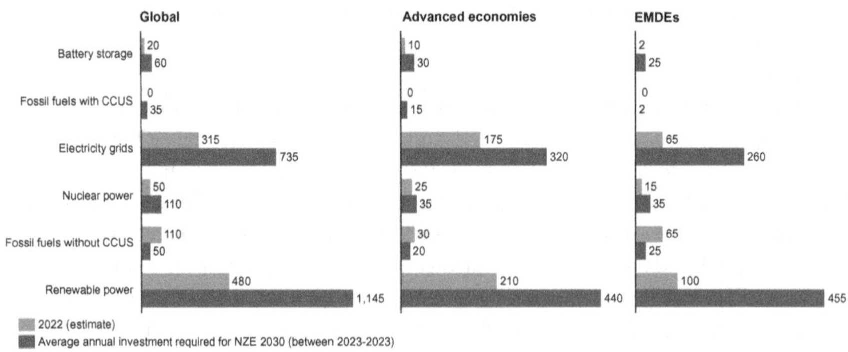

Fig. 6.23 Investment in the power sector in 2022 versus average annual investment required between 2023 and 2030 for NZE 2050

other renewables. Lastly, small hydro faces modestly declining investment growth and relatively low investment volumes.

Finally, in order to reach net zero emissions 2050 target, major increase is needed in climate technology investments and technologies that enable the phasing out of fossil fuels. Based on IEA World Energy Investment data, we are far behind in all technology platforms (IEA WEI 2022) (Fig. 6.23).

References

Bloomberg (2022) Bloomberg NEF annual reports. https://about.bnef.com/new-energy-outlook/

Climate Bonds Initiative (2022) https://www.climatebonds.net/market/data/

Climate Policy Initiative (2022) Climate policy initiative energy database. https://climatepolicydatabase.org/

EIB (2022) European investment bank group sustainability report 2022. https://www.eib.org/en/publications/20230023-sustainability-report-2022

GlobalData (2022) https://www.globaldata.com/industries-we-cover/globaldata-disruptor/

GlobalData (2023) GlobalData disruptor platform, deals database. https://www.globaldata.com/data/

ICCT (2022) Electrification of heavy-duty vehicles in emerging markets. https://theicct.org/publication/global-hvs-evs-zev-electrif-hdv-emerg-mkts-sep22/

IEA (2022) Wind electricity. IEA, Paris. https://www.iea.org/reports/wind-electricity

IEA EV (2022) Public EV charging stations, dataset. Electric bus fleet by country, dataset. https://www.iea.org/reports/global-ev-outlook-2022/trends-in-charging-infrastructure. https://www.iea.org/reports/electric-vehicles

IEA GET (2022) Governmental Energy Technology R&D country budgets, database and Global Hydrogen Review. https://www.iea.org/reports/global-hydrogen-review-2022

IEA WEI (2022) World energy investment. Planned electrolyser manufacturing capacity by region, dataset. https://www.iea.org/reports/world-energy-investment-2022. https://www.iea.org/data-and-statistics/charts/planned-electrolyser-manufacturing-capacity-by-region-2021-2030

IEA WEO (2022) World energy outlook 2022. https://www.iea.org/reports/world-energy-outlook-2022

IFC (2022) https://www.ifc.org/wps/wcm/connect/corp_ext_content/ifc_external_corporate_site/about+ifc_new/investor+relations/ir-products/grnbond-overvw

OECD (2022) Aggregate trends of climate finance provided and mobilised by developed countries in 2013–2020. https://www.oecd.org/climate-change/finance-usd-100-billion-goal

OECD (2022) Climate change mitigation patents database. https://data.oecd.org/envpolicy/patents-on-environment-technologies.htm

Schoenmaker D, Schramade W (2019) Principles of sustainable finance. Oxford University Press. ISBN 978-0-19-882660-6

Statista (2021) https://www.statista.com/study/51447/hydrogen/

UNEP (2020) 2019. Frankfurt School-UNEP Centre, BloombergNEF. Global trends in renewable energy investment. https://www.fs-unep-centre.org/global-trends-in-renewable-energy-investment-2020/

WBG (2023) https://www.worldbank.org/en/home

WEF (2022) Review 57(2):66–90. https://www.weforum.org/agenda/2022/11/climate-change-climate-adaptation-private-sector/Management. https://doi.org/10.1525/cmr.2015.57.2.66

Chapter 7
Carbon Neutrality and Entrepreneurship

Abstract Start-ups are playing a crucial role in driving climate technology innovation within the global innovation ecosystem. We discuss both outside-in and inside-out innovation models that are highly relevant to climate technology, allowing corporations to benefit from external innovations. For instance, start-ups actively contribute to AI solutions for achieving a carbon-neutral society and big data analysis in the Energy Internet (EI). In areas like direct air capture and green hydrogen concepts, start-ups leverage scientific knowledge, academic collaborations, and innovative partnerships, placing them in a strategic position within the climate technology space. Many start-ups in this sector have strong connections to academia, with advisory board members or founders from renowned research institutions, facilitating collaborations with universities. In the past decade, the total value of climate technology first-round venture capital financing deals has notably increased, reflecting the growing importance of start-ups in developing emerging climate technologies. As many technologies are needed to meet ambitious 2050 targets, they are still in the demonstration or prototype phase, start-ups play a crucial role in driving the development of these innovations. Their contributions are essential in advancing sustainable solutions and accelerating the journey towards a greener future.

Keywords Entrepreneurship · Innovation ecosystem · Start-up companies · Venture capital · Climate Technology Start-ups · Climate Change mitigation

7.1 Basics

Addressing climate change mitigation requires a substantial level of innovation across various domains, including technology development, policy implementation, finance, and the establishment of new business models. The global innovation ecosystem predominantly relies on large corporations, prominent research institutions like national laboratories and academia. Start-up companies have also been in the very core of innovation, and many of the current technology giants in the ICT

M. Hakovirta, *Carbon Neutrality*, Springer Climate,
https://doi.org/10.1007/978-3-031-45202-4_7

sector are a good example of that. Their groundwork was not just based on new business model creation but also on technology advancement created by talented early career individuals with entrepreneurial mindset. In terms of climate change, the role of entrepreneurship and start-up businesses can also be significant in accelerating and driving climate innovation. The market demand for these new businesses is timely as fact-based evidence of climate change, and severe weather events exist, and the world is collectively looking for solutions. The overall concern of global citizens has created a strong consumer awareness and demand.

The latest global policy changes and the general trend in corporate climate commitments with highly challenging targets have created many business growth opportunities. It is evident that for large corporations there are challenges in the understanding of what the necessary solutions for reaching their targets mean. Many of these developments demand novel and agile corporate strategies and new partnerships for their implementation. Some of these solutions will also include major changes in new business models including transitioning to bioeconomy, circular economy, and becoming familiar with such concepts as Biomimicry and natural capital (Hakovirta 2022).

In many cases, the innovativeness of a company is not solely determined by its products or services but rather by its capability to align its strategies with market and customer demands. Therefore, the climate change and carbon zero targets have become an important part of companies' business strategies. The journey towards carbon neutrality is a lengthy one, and it poses particular challenges for industries with high GHG emissions. These industries are also often characterized as mature and not heavily reliant on R&D and their industry renewal and overall speed of innovation is slow. As an example, many of the power plants that are built, last as long as 30–50 years with a very long ROI. The advancement of climate technology often necessitates long-term investments in not only R&D but also in acquiring new skills and capabilities. Furthermore, the development of climate technology heavily relies on government policies, which are often regionally or locally oriented. The market prices of products may also be highly volatile and create a need for high business risk tolerance. The biofuels industry in the early 2000s serves as a notable example where government subsidies created an artificial level of competitiveness for ethanol and biodiesel production in comparison with their fossil fuel-based counterparts.

In this context, the development of sustainability and innovation strategies is increasingly crucial. Companies need to use different *innovation management models* to strengthen the efficiency and effectiveness of their innovation processes (Barbaroux et al. 2016). Innovative companies often leverage external information sources, such as patent landscaping, actively scouting for new ideas, and establishing value-adding networks and partnerships with start-up companies (Boston Consulting 2018). Collaborating with leading start-ups can assist large corporations in driving innovation, accelerating time to market, and establishing business ecosystems. Such partnerships can facilitate the integration into new value chains and expedite the scaling of technologies, often surpassing the capabilities of relying solely on their internal R&D organizations (Prashantham 2021).

Increasingly start-up companies have become part of innovative corporations' innovation strategies. Moreover, partnering with start-ups can attract investors who recognize the significant potential of new business opportunities and the fresh entrepreneurial talent that can invigorate innovation. This enables companies to secure the necessary resources and execute its strategic initiatives more effectively (Ghezzi et al. 2022). Entrepreneurship and start-up activities have emerged as integral components of competitive advantage in driving innovation for climate change solutions. The dynamic and agile nature of start-ups, coupled with their willingness to take risks and explore new ideas, has positioned them as key players in developing and implementing innovative solutions to address climate change challenges. Their ability to adapt quickly, leverage emerging technologies, and challenge traditional approaches brings a fresh perspective to the pursuit of sustainable solutions. Consequently, entrepreneurship and start-up activities play a vital role in fostering innovation and advancing the fight against climate change.

7.2 Innovation Ecosystem and Start-Up Companies

Corporations are constantly engaged in the evolution and refinement of their products, services, and processes in order to establish future business growth and stability. However, the major challenge lies in effectively managing the delicate balance between short-term and long-term value creation, particularly within mature businesses. Striking this equilibrium requires careful attention to simultaneously meeting immediate stakeholder expectations while investing in innovation, sustainability, and long-term viability. It calls for strategic decision making and a comprehensive approach that integrates immediate profitability with sustainable, enduring growth. Similarly, corporate goals of mitigating GHG emissions entail a delicate balancing act of maximizing returns on previous capital investments while ensuring a sustainable future through new technology platforms and carbon–neutral business models.

In this context, organizations and leaders must adopt an ambidextrous approach, consecutively exploring new opportunities and diligently leveraging existing capabilities. This requires a mindset that embraces both exploration and exploitation, allowing for the pursuit of innovative solutions while effectively utilizing current resources and expertise. By striking this balance, organizations can effectively navigate the transition towards GHG emission mitigation and sustainable practices (Carlos et al. 2021). Since past performance can predict future direction, this task is especially challenging. Companies relying solely on internal resources and fixed capabilities often face challenges in adapting their strategies and implementing them effectively. Conversely, one of the key strengths of start-up companies, especially in their early stages, is their ability to align strategy, structure, competencies, and capabilities. This grants them the necessary flexibility to operate in unfamiliar business environments, offering greater degrees of freedom for experimentation and adaptation. By having the agility to align all these elements from the outset, start-ups can better navigate uncertainties and seize opportunities in dynamic and rapidly evolving markets (Hakovirta et al. 2022).

Manufacturing-centric corporations that are large, mature, and in a stable evolutionary phase often have energy-intensive operations that contribute to GHG emissions. Consequently, these companies find themselves at the forefront of GHG mitigation efforts and science-based target setting. Given their significant environmental footprint, these corporations play a crucial role in addressing GHG emissions and transitioning towards more sustainable practices. By setting science-based targets, they establish clear objectives grounded in scientific evidence, which guide their efforts to reduce emissions and contribute to global climate goals. Their actions and initiatives can have a substantial impact on overall GHG mitigation efforts and serve as important examples for other industries to follow. As earlier discussed, the most GHG emitting sectors are in energy (electricity and heat production), chemical, metallurgical, and mineral industries (with fossil fuels used in on-site operations), agriculture, forestry and land use sectors, transportation (road, rail, air, and marine transportation), construction and building (Circularity Gap Report 2022; IPCC 2014). It is true that sectors characterized as less innovative face challenges in climate innovation. This highlights the potential significance of start-up companies in helping achieve the science-based targets that corporations in these sectors have committed to. Start-ups, with their agility and fresh perspectives, can bring innovative solutions to the table and assist in driving the necessary transformations. By collaborating with start-ups, established corporations can tap into external sources of innovation, leverage emerging technologies, and access entrepreneurial talent that can accelerate progress towards their climate goals. Such partnerships can facilitate the adoption of new technologies, business models, and practices that aid in meeting science-based targets and promoting sustainability (Bolton and Kacperczyk 2021).

In general, corporations adopt two models for engaging with start-up companies. The first model is known as outside-in innovation, which involves making the technology of existing start-ups accessible and finding a valuable fit for specific corporate needs. This approach allows corporations to leverage external innovations and incorporate them into their operations, products, or services to enhance competitiveness and address specific challenges.

The second model is the inside-out open innovation approach, where the focus is on finding uses for the corporation's own technology in other companies or industries. In this case, the corporation aims to identify new applications, markets, or partnerships where its technology can create value beyond its current scope. This approach enables the corporation to explore diverse opportunities and generate additional value by sharing or licensing its technology to external entities. Both models offer distinct pathways for collaboration between corporations and start-ups, allowing for the exchange of knowledge, resources, and capabilities to drive innovation and create mutual benefits (Weiblen and Chesbrough 2015).

Both the outside-in and inside-out innovation models are highly relevant to climate technology, as they enable corporations to benefit from external innovations while also diffusing capital-intensive technology solutions needed for achieving carbon neutrality. Leveraging external innovation allows corporations to identify market-aligned business opportunities in a timely manner. At the same time, incorporating start-up developments into their strategies enables them to orchestrate value chains and develop innovative business models.

It is worth noting that for many companies, making substantial investments in start-up acquisitions or managing partial ownership can pose challenges. Conversely, relying solely on large in-house R&D organizations and research facilities can be highly expensive. In certain cases, it can be more cost-effective for large corporations to tap into start-up developments and simultaneously perform value chain orchestration and business model development. This allows them to leverage the existing partnerships, customers base and business expertise of start-ups while avoiding excessive costs associated with internal research and development.

By engaging with start-ups, corporations can access cutting-edge technology, entrepreneurial talent, and novel business ideas, positioning themselves at the forefront of climate technology innovation. This approach facilitates collaboration, knowledge exchange, and the exploration of new market opportunities, contributing to both the corporation's sustainability goals and its long-term business success.

As for example, the active engagement of the start-up community is noteworthy in the development of AI solutions for achieving a carbon–neutral society and the utilization of big data analysis in the Energy Internet (EI). The EI aims to significantly enhance smart grids by integrating green energy platforms into a flexible and efficient grid infrastructure. Start-ups have played a pivotal role in driving innovation in these areas. Moreover, they have harnessed the power of big data analysis to optimize the performance of the EI, enabling the integration of diverse green energy sources into a resilient and effective grid system. The collaboration between start-ups and the EI sector has yielded remarkable advancements, propelling the transformation of traditional energy grids into intelligent and sustainable networks. This collaboration underscores the significance of entrepreneurial ingenuity and innovative thinking in accelerating progress towards a carbon–neutral future (Kabalchi et al. 2019).

Examples of advancements by companies also include direct air capture (DAC) technology development by for example Carbon Engineering Ltd., Global Thermostat LLC, and Climeworks AG). These companies have successfully commercialized a groundbreaking process that involves capturing CO_2 from the atmosphere and converting it into low-carbon fuels using renewable energy sources. They serve as excellent examples of pioneers in developing carbon–neutral solutions within a challenging market environment.

The unique aspect of their work is that they have managed to navigate the market landscape despite the absence of a well-established business model, apart from the sale of carbon credits. Additionally, the market for the extracted CO_2 is currently limited. Nonetheless, these companies have demonstrated their ability to innovate and create sustainable solutions, even in the face of such challenges. Their achievements highlight their visionary approach and commitment to addressing climate change by offering practical and scalable alternatives in the form of low-carbon fuels. By leveraging renewable energy sources and employing advanced carbon capture technologies, they are at the forefront of transforming CO_2 into valuable resources while minimizing environmental impact.

The primary objective of these start-ups, or similar ventures, is to develop the capability to produce green hydrogen and combine it with captured CO_2 to create low-carbon fuels for medium and heavy transportation vehicles worldwide (Boretti 2013; Roh et al. 2015). A notable characteristic of start-ups in the climate technology sector is their strong alignment with scientific principles and their close connections to academia and research institutions. In a recent study by Capgemini (Gapgemini 2020), it was found that corporations are increasingly recognizing the value of external innovation, leading to the growing popularity of universities, crowdsourcing, and start-ups as sources of fresh ideas and advancements. This trend reflects a shift towards embracing collaborative approaches and tapping into the expertise of external entities for innovation and problem-solving.

By leveraging scientific knowledge, academic collaborations, and innovative partnerships, start-ups in the climate technology space are well-positioned. Their commitment to a science-based approach enables them to develop cutting-edge solutions that address pressing environmental challenges and create a positive impact on society. Start-ups in the climate technology sector often benefit from strong academic connections and networks, with advisory board members or even founders who come from academia or renowned research institutions. This integration of academia into start-up ventures underscores the significance of start-ups in driving the development of climate technology solutions.

For large corporations, collaborating with academia on market-focused innovations can be challenging due to non-disclosure or confidentiality constraints. This perspective on start-up collaboration is equally important for SMEs, which typically lack direct connections to academia due to the associated collaboration costs and the need for rapid product development cycles. In addition, the lack of scientific and R&D managerial skills can create additional challenges. Despite these obstacles, start-ups play a crucial role in innovation ecosystems by facilitating the transfer of scientific intellectual property into viable business perspectives. Start-ups act as vital bridges between scientific knowledge and practical applications, and their ability to bridge the gap between academia and business contributes to the overall innovation landscape and enhances the commercialization of scientific breakthroughs in the climate technology sector.

7.3 Climate Technology and Start-Ups

PitchBook is a comprehensive database that specializes in mergers and acquisitions, private equity, and venture capital activities. It gathers information from a wide range of sources, including over 3 million private companies, over 100,000 publicly traded companies, and companies that have received funding from venture capital and private equity firms. The database also includes data on more than 50,000 start-ups, stealth start-ups (those operating in stealth mode), and numerous pre-IPO companies.

With its extensive coverage of both private and public companies, as well as its focus on investment and financial activities, PitchBook provides valuable insights and data for professionals in the finance, investment, and entrepreneurial sectors. The database serves as a valuable resource for tracking market trends, conducting industry research, and identifying potential investment opportunities across various stages of a company's lifecycle.

In connection with this, start-up companies are typically categorized as formative or early-stage companies that receive backing from venture capital firms. These companies are in their initial phases of development and are often characterized by high levels of innovation and growth potential.

Early-stage start-ups commonly secure a round of financing within the first five years of their existence, which helps fuel their growth and expansion. Once a start-up has successfully received funding and progresses beyond the initial stages, it is often classified as a later-stage company. The distinction between early-stage and later-stage start-ups is based on the company's maturity, growth trajectory, and funding stage. Early-stage start-ups are typically focused on product development, market validation, and establishing a solid foundation for future growth. Later-stage start-ups, on the other hand, have typically advanced beyond these initial stages and are more focused on scaling operations, expanding market reach, and achieving profitability. The categorization of start-ups into different stages provides insights into their current development phase and funding status, allowing investors and stakeholders to understand the level of risk and growth potential associated with these companies.

In terms of data analysis, industry verticals are used and refer to a clearly defined group of companies that operate within a specific market or industry. These companies often share similar characteristics and target a common customer segment. However, an industry vertical may also encompass companies that overlap with multiple industries, sharing common objectives, or addressing similar challenges.

The industry vertical "Climate Tech," comprises of companies that offer or develop products, services, and technologies aimed at mitigating climate change. These companies focus on creating innovative solutions to reduce greenhouse gas (GHG) emissions and combat the effects of climate change. Within the Climate Tech vertical, there are diverse subcategories that encompass various sectors. For example, companies may specialize in electrifying industrial processes and transportation to reduce carbon emissions, develop agricultural innovations that promote lower GHG emissions, or create low-carbon mining technologies. These are just a few examples of the breadth of solutions that fall within the Climate Tech vertical.

By categorizing companies within industry verticals like Climate Tech, it becomes easier to analyze market trends, track innovations, and identify investment opportunities within the specific sector. This vertical approach helps stakeholders gain a better understanding of the companies operating in the field of climate technology and fosters collaboration and knowledge sharing within the industry.

7.4 Progress and Role of Climate Technology Start-Ups

Looking at the *climate technology vertical* data from 2000 to 2021, the VC funding has taken a major uptake. When comparing the number of climate technology start-ups to the total number of start-ups, there has been a remarkable 20-year change, witnessing a significant increase of 2000%. In the past, climate technology start-ups accounted for only 0.1% of the total start-up landscape, but this has risen to 2.1% in recent years. This indicates a substantial growth and heightened focus on developing innovative solutions to address climate change challenges.

A similar trend can be observed when examining the share of financial investments or deals in the climate technology sector. There has been an impressive relative growth of 2811% in deal counts over the same 20-year period. Initially, climate technology investments accounted for a mere 0.09% of all investments, but this has surged to 2.62% in recent times.

These statistics demonstrate the increasing prominence and momentum of the climate technology sector within the broader start-up and investment landscape. The substantial growth in both the number of climate technology start-ups and the share of financial investments underscores the growing commitment to addressing climate change and the urgent need for innovative solutions (Fig. 7.1).

Looking at more closely at the data, it indicates that environmental sustainability was already gaining traction and making significant strides in the start-up world as early as the 2000s. This suggests that concerns about the environment and the need for sustainable solutions were becoming more prominent in the entrepreneurial landscape.

A notable surge in climate technology start-ups can be observed from 2013 onwards, indicating a concentrated effort to address climate change and develop innovative solutions during this period. This could be attributed to various factors, such as increasing awareness of environmental issues, policy changes, and advancements in technology that enabled the emergence of new climate-focused ventures.

However, it's important to note that the data also reveals a decline in the number of climate technology start-ups starting in 2020, which can largely be attributed to the global COVID-19 pandemic. The pandemic has had a profound impact on the global economy, including the start-up ecosystem. Many companies faced financial challenges, operational disruptions, and shifting priorities, which likely affected the growth and establishment of new climate technology ventures. Despite the temporary setback caused by the pandemic, the overall upward trend in climate technology start-ups and investments highlights the long-term commitment to sustainability and the recognition of its importance in addressing environmental challenges.

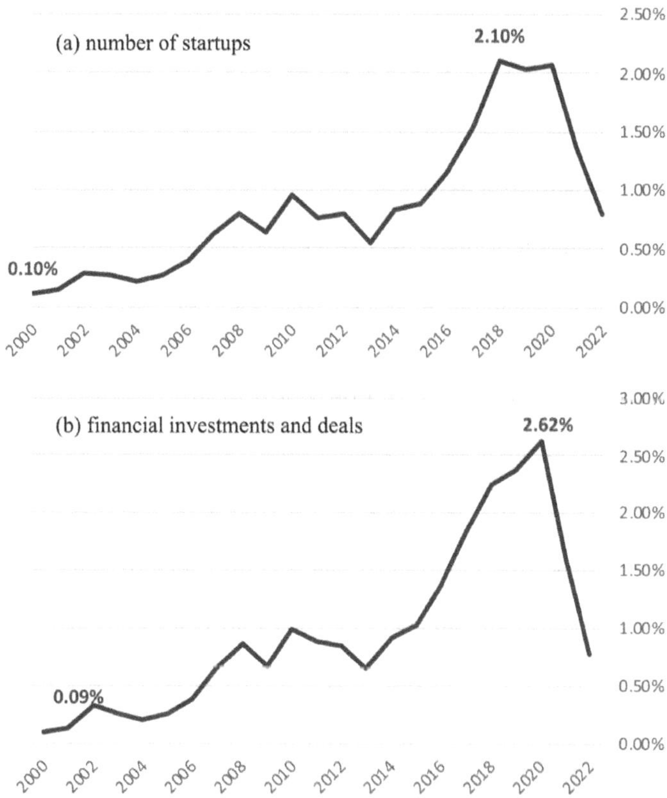

Fig. 7.1 Share (%) of climate technology start-ups to all-start-ups **a** number of companies and **b** number of deals (2000–2022) (Hakovirta et al. 2022)

7.5 Industry Segments and Climate Technology

If we analyze the 20-year period from 2000 to 2021, the majority of climate technology start-ups can be categorized into various industry segments. The distribution of these start-ups across industry segments is as follows:

Food System: Approximately 24% of climate technology start-ups are focused on the food system. These companies aim to develop sustainable solutions related to agriculture, farming practices, food production, and distribution.

Energy Grid Technology: Around 18% of climate technology start-ups are engaged in the development of energy grid technologies. These companies focus on innovations related to the electrical grid, smart grid systems, energy storage, and grid optimization.

Green Energy Generation: Roughly 16% of climate technology start-ups are dedicated to green energy generation. These companies work on renewable energy sources such as solar, wind, hydro, and geothermal power, aiming to increase the share of clean energy in the overall energy mix.

Electric Transportation: Approximately 12% of climate technology start-ups focus on electric transportation. These companies develop electric vehicles (EVs), charging infrastructure, and related technologies to promote sustainable and zero emission transportation solutions.

Mobile Solutions (Enhanced Mobility using Digital Devices and Services): Around 10% of climate technology start-ups are involved in the development of mobile solutions that enhance mobility using digital devices and services. These companies leverage technology to provide innovative transportation options, shared mobility, and intelligent transportation systems.

Land Use: Approximately 6% of climate technology start-ups are focused on land use. These companies work on sustainable land management, reforestation, ecosystem restoration, and land conservation to mitigate the environmental impacts of land use practices.

Built Environment: Roughly 6% of climate technology start-ups concentrate on the built environment. These companies develop sustainable building materials, energy-efficient construction practices, and smart building technologies to reduce the environmental footprint of the built environment.

Industry: Around 5% of climate technology start-ups are involved in industrial solutions. These companies focus on developing technologies and processes that promote energy efficiency, emissions reduction, and sustainable practices within various industries.

Carbon Technology: Approximately 4% of climate technology start-ups are dedicated to carbon technology. These companies work on carbon capture, utilization, and storage (CCUS) technologies, as well as other carbon management solutions to mitigate greenhouse gas emissions.

These industry segments reflect the diverse areas where climate technology start-ups are making significant contributions, addressing various aspects of climate change mitigation and sustainability. Remarkably there is a clear connection between the industry segments of climate technology start-ups and the earlier introduced societal needs-based categorization (Circularity Report 2022). For example, the food system segment, which accounts for a significant portion of climate technology start-ups, aligns with the societal need for addressing the environmental impact of the food industry. The food industry value chain is known to contribute a substantial amount of greenhouse gas (GHG) emissions, making it a key area of focus for climate change mitigation.

According to the Circularity Gap Report of 2022, the food industry value chain is responsible for approximately 16.9% of global GHG emissions, amounting to around 10.0 billion tons. This makes it the third-largest contributor to climate change after the energy sector and industrial processes. By addressing sustainability challenges in the food system, climate technology start-ups can play a crucial role in reducing emissions and promoting more sustainable practices.

Other similar alignment can be observed in transportation, which for example accounts for a significant portion of GHG emissions, approximately 16.2% of the total emissions according to your information. This aligns with the presence of climate technology start-ups in the electric transportation segment, indicating their efforts to address emissions in the transportation industry.

Similarly, the forestry and land use operations, which contribute around 11% of GHG emissions according to the IPCC report, also overlap with the industry segments of climate technology start-ups. This highlights the role of start-ups in developing innovative solutions for sustainable forestry practices, land management, and carbon sequestration, aiming to reduce emissions and promote sustainable land use.

The correlation between the industry sectors of climate technology start-ups and the highest GHG emitting activities and sectors indicates a strategic focus on addressing critical emission sources and areas with the greatest potential for impact. By targeting these sectors, start-ups can contribute to significant emissions reductions and play a crucial role in addressing the urgent challenges of climate change and global environmental crises. This insight emphasizes the important role of start-up businesses within the global innovation ecosystem, as they actively develop and advance solutions for climate change mitigation.

Another noteworthy factor is to see the latest numbers of the distribution of *first-round VC financing* by industrial segment. In 2021, energy transition–grid technology was the highest (50%), general industry category was the second largest (14%) and the land use was in the third place (11%) followed by a 10% share by food systems. Interestingly, electric transportation, clean energy, carbon technology, and built environment are in a lagging role far behind the other segments (Fig. 7.2).

Premoney and post-money valuations play a crucial role in understanding the financial aspects of start-up investments. The premoney valuation refers to the estimated value of a company before any external investment or financing is obtained. It serves as a basis for determining the ownership stake that investors will receive

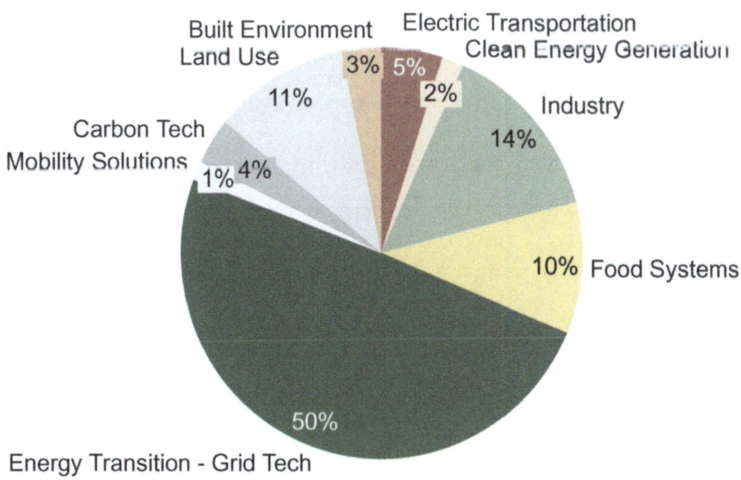

Fig. 7.2 Share of total first-round VC financing by segment in 2021

in exchange for their investment. However, premoney valuations are subjective and can vary depending on various factors (Köhn 2018).

The assessment of premoney valuations involves analyzing the start-up's strategy, financials, management team, advisory board, and comparable exits. This evaluation is typically conducted by assessors who consider multiple aspects to estimate the worth of the company at that point in time (Copeland et al. 2000).

On the other hand, the post-money valuation builds upon the premoney valuation and includes the amount of financing raised. It represents the total value of the company after the investment or financing has been secured. In simple terms, it is the sum of the premoney valuation and the investment amount.

Understanding premoney and post-money valuations is essential for investors, as it helps determine the value and potential return on investment. These valuations provide insights into the financial health and growth prospects of start-up companies, including those in the climate innovation space.

7.6 Significance of Start-Ups in Climate Change Mitigation

If we look more closely at the past 10 years, the total value of climate technology first-round VC financing deals has been also increasing considerably. Although COVID-19 did impact the numbers, they are on a fast-paced trajectory to reach the 2018 level. The share of climate technology VC deals out of all deals is however decreasing (Fig. 7.3).

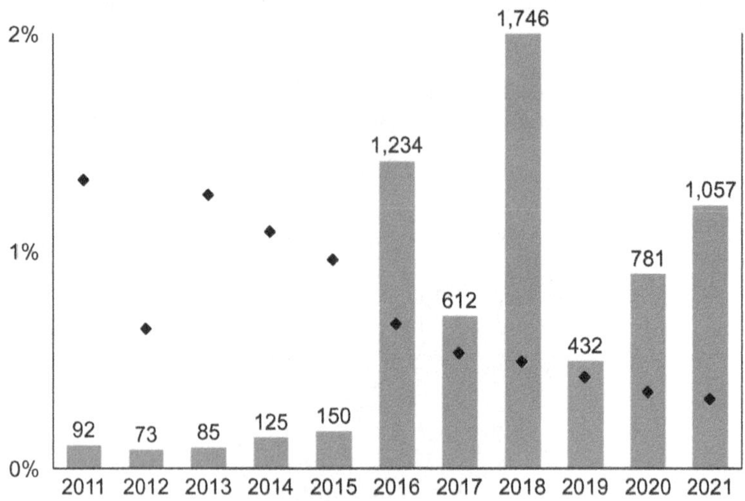

Fig. 7.3 Value of first-round VC deals (in USD million) in climate technology and the share of climate tech VC deals out of all deals, 2011–2021

In general, the USA is considered as the most start-up friendly country and this can be seen looking at the climate technology vertical as well by looking at both company and deal counts. Europe and Asia, particularly China and the UK, are the other prominent regions when it comes to climate technology. These regions have shown significant commitment and progress in addressing climate change and fostering innovation in clean technologies (Fig. 7.4).

The higher number of climate technology start-ups in China can be attributed to the Chinese government's strong emphasis on clean technology and environmental remediation. In recent years, the Chinese government has implemented policies and initiatives to promote the development and adoption of cleantech solutions, recognizing the importance of addressing environmental challenges and reducing greenhouse gas emissions. This has created a favorable environment for the growth of climate technology start-ups in China.

The significant peak in climate technology start-ups in 2014 may be linked to various factors. The introduction of the EU Directive 2014/95, which required large

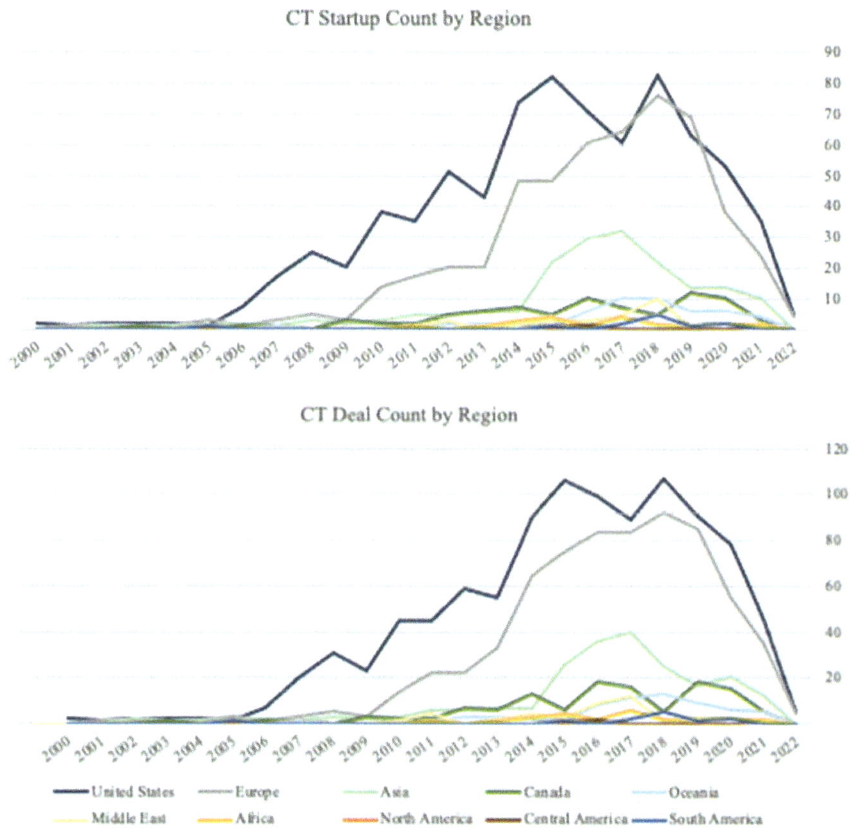

Fig. 7.4 Number of climate innovation start-ups and deals by region (Hakovirta et al. 2022)

companies to disclose environmental information, likely increased the focus on environmental matters and encouraged more activity and investments in climate technology. Additionally, the launch of the United States Climate Action Report in 2014 and the development of the UN Sustainable Development Goals (SDGs) created a heightened awareness and accountability around climate issues, potentially driving increased interest and investment in climate technology during that period.

The year 2018 marked a significant development in the trade relationship between the USA and China. The USA initiated the imposition of tariffs and trade barriers on Chinese goods, citing concerns of unfair trade practices and intellectual property theft. This trade dispute may have had some impact on the climate technology landscape, potentially affecting the flow of investments and collaborations between the two countries in this sector. Overall, the combination of government initiatives, regulatory changes, increased awareness of environmental issues, and geopolitical dynamics has influenced the growth and activity of climate technology start-ups in different regions, including China.

There was a surge in climate technology start-up investments observed in the USA, Europe, and Canada during 2017–2018. This could have been influenced by the release of the IPCC Special Report on the impacts of global warming of 1.5 °C. This report, which highlighted the urgent need for climate action, increased global awareness and drew attention to the importance of investing in climate solutions. The investment community, recognizing the potential market and impact of climate technology start-ups, may have responded by increasing their investments in this sector.

Also, during 2018 natural disasters such as Hurricane Florence in the USA and forest fires in California, British Columbia, and northern Europe had devastating impacts. These events served as wake-up calls, highlighting the urgent need to address climate change and its consequences. They likely increased public awareness and amplified the demand for climate solutions, creating a favorable market for climate technology start-ups. The presentation of the European Green Deal in the following year further accelerated investments in climate solutions in Europe. This comprehensive plan outlined ambitious goals and measures to address climate change and promote sustainability, signaling a strong commitment from the European Union. The European Green Deal likely contributed to the growing interest and investments in climate technology start-ups within the region.

Start-ups are playing an increasingly vital role in the global innovation ecosystem, particularly in driving climate technology innovation. The current pace of innovation by traditional corporate strategies is often considered too slow to effectively address climate change. In contrast, start-up businesses have the potential to accelerate innovation and drive societal change in environmental sustainability and climate solutions.

Recent reports, such as the IEA net zero by 2050 report, emphasize the importance of deploying the best available technologies today to achieve significant CO_2 emissions reductions by 2030. However, many of the technologies required to meet the ambitious targets for 2050 are still in the demonstration or prototype phase. This highlights the significance of start-ups in developing emerging climate technologies.

The increasing flow of deals and capital investments into start-ups in the climate technology sector indicates the growing importance of these companies. Corporations should recognize the value of connecting with start-ups and incorporate them explicitly into their strategic plans for achieving carbon neutrality.

Engaging with start-ups offers several benefits, including accelerated speed of innovation, greater operational flexibility in the climate technology business environment, and more effective collaboration with academia. Moreover, it enables the efficient transfer of scientific intellectual property to business applications. When well-orchestrated, these factors can provide a clear competitive advantage to corporations.

In summary, start-ups are critical drivers of climate technology development, and their integration into corporate strategies can lead to enhanced competitiveness and facilitate the transition to a sustainable future.

References

Barbaroux P, Attour A, Schenk E (2016) Innovation processes, innovation capabilities and knowledge management. In knowledge management and innovation (eds P. Barbaroux, A. Attour and E. Schenk). https://doi.org/10.1002/9781119330134.ch1

Bolton P, Kacperczyk M (2021) Firm commitments, Columbia Business School Research Paper. https://doi.org/10.2139/ssrn.3840813

Boretti A (2013) Renewable hydrogen to recycle CO_2 to methanol. Int J Hydrogen Energy 38(4):1806–1812. https://doi.org/10.1016/j.ijhydene.2012.11.097

Boston Consulting (2018) Most innovative companies 2018: innovation and digital. https://www.bcg.com. Boston Consulting Group, 17 Jan 2018, www.bcg.com/publications/collections/most-innovative-companies-2018.aspx

Carlos A, Alvares T, Barbieri JC, Carlos de Morais DO (2021) Horizontal innovation and ambidextrous organization: a new innovation model applied in a mature industrial company. Int J Innov 9(3):588–621. https://doi.org/10.5585/iji.v9i3.19012

Circularity Gap Report (2022) https://www.circularity-gap.world/2022#Download-the-report. 15 Mar 2022

Copeland T, Koller T, Murin J (2000) Valuation: measuring and managing the value of companies. Wiley, New York

Gapgemini (2020) Lifting the lid on corporate innovation in the digital age, 12 May 2020. Retrieved from https://www.capgemini.com/wp-content/uploads/2020/05/MIT-INVENTReport_NEW-2020.pdf

Ghezzi A, Cavallo A, Sanasi S, Rangone A (2022) Opening up to startup collaborations: open business models and value co-creation in SMEs. Compet Rev 32(7):40–61. https://doi.org/10.1108/CR-04-2020-0057

Hakovirta M (2022) Strategic sustainability management, ISBN 9781792460807. Kendall hunt publishing company

Hakovirta M, Kovanen K, Martikainen S, Manninen J, Harlin A (2022) Corporate net zero strategy—opportunities in start-up driven climate innovation. In: Business strategy and the environment, pp 1–12. https://doi.org/10.1002/bse.3291

IPCC (2014) Pachauri RK, Meyer LA (eds) Climate change 2014: synthesis report. Contribution of working groups I, II and III to the fifth assessment report of the intergovernmental panel on climate change. Core writing team, IPCC, p 151

Kabalci Y, Kabalci E, Padmanaban S, Holm-Nielsen JB, Blaabjerg F (2019) Internet of Things applications as energy internet in smart grids and smart environments. Electronics 8(9):972. https://doi.org/10.3390/electronics8090972

Köhn A (2018) The determinants of startup valuation in the venture capital context: a systematic review and avenues for future research. Manag Rev Q 68:3–36. https://doi.org/10.1007/s11301-017-0131-5

Prashantham S (2021) Partnering with startups globally: distinct strategies for different locations. Calif Manage Rev 63(4):123–145. https://doi.org/10.1177/00081256211022743

Roh K, Nguyen TBH, Suriyapraphadilok U, Lee JH, Gani R (2015) Development of sustainable CO_2 conversion processes for the methanol production. Comput Aided Chem Eng 37:1145–1150

Weiblen T, Chesbrough HW (2015) Engaging with startups to enhance corporate innovation, California

Chapter 8
Socioeconomic Aspects of Climate Change in Cities and Municipalities

Abstract In this section, we discuss how climate change profoundly affects social structures, particularly in rural and underprivileged coastal areas, leading to devastating impacts from extreme weather events like floods, storms, and droughts. Urban areas also suffer from rising sea levels, disruptions to infrastructure and services, job losses, and health and well-being challenges. The majority of vulnerable megacities are situated in coastal areas, highlighting the need for global cooperation and support in addressing climate change fairly and equitably. Developing countries in Asia and Africa experience significant environmental issues due to the global trend of urbanization, including pollution, poverty, and overcrowding. Pacific Islands are threatened by ocean acidification impacting fisheries and coral reefs. Developing countries must receive necessary assistance and resources to cope with the challenges they face in mitigation and adaptation. Addressing environmental or climate justice is crucial in developing global climate solutions, with a focus on fair distribution of resources and opportunities. Climate change profoundly affects traditional industries, retail, commercial services, tourism, insurance, and agriculture. Mitigation efforts in cities and municipalities are critical, with sustainable cities and eco-cities playing an essential role in the equation. Overall, a cultural change and mindset adjustment is crucial in addressing climate change. Communities that value sustainability and environmental responsibility can motivate individuals to take actions to mitigate climate change.

Keywords Socio-economical impact · Social impacts of climate change · Developing countries · Island nations · Coastal cities and metropolitan areas · Mitigation efforts in cities and municipalities

8.1 Basics

Climate change is a phenomenon that impacts social structures in all its forms. Rural and many times poor or modest-income underprivileged coastal areas suffer tremendously from the volatility of the climate change effect. The global impact is many

M. Hakovirta, *Carbon Neutrality*, Springer Climate,
https://doi.org/10.1007/978-3-031-45202-4_8

times highly intertwined to localities with urban life including rising sea level, due to extreme weather events such as floods, storms, and droughts. These and other related effects create disruption to urban life including basic infrastructure and services, jobs, health, and well-being. As urban growth has been part of industrialization for the past two centuries, we now have more people living in urban areas than in countryside. It has been estimated that by 2030, 5 billion people will live in towns and cities.

The change is exponential, and already by 2050 more than 70% of the global population will live in urban environment (Hakovirta 2021). For these reasons, it is no surprise cities and towns, and their built environment are the main contributors to climate change, and with most of the global population living in them, they are a major source for greenhouse gas emissions. Transportation and the built environment including buildings and housing are estimated to create over 70% of the CO_2 emissions (Wei et al. 2021). The global megatrend of urbanization is largely impacting developing countries in Asia and Africa and is creating major issues in pollution, poverty, and overcrowding. Such changes have also produced major economic, social equity, and segregation issues. In these and other areas of high social vulnerability and poverty exist, major actions are needed for climate change. These actions are necessary for the climate change-induced social crisis, including inequality that is becoming more polarized globally. *Environmental or climate justice* needs to be addressed in developing global climate solutions including principles of *distributive justice*; the recognized fairness of how people judge what they receive. The developed countries are already investing in climate and adaptation actions such as coastal resiliency programs, development or green urban and city infrastructure, and sustainable energy solutions. Ideally, these test platforms and developed climate solutions can be later adopted or transferred to developing countries with the support of global initiatives by governments, corporations, and institutional financial entities.

8.2 Social Impacts of Climate Change

8.2.1 Developing Countries

Climate change has relatively recently been seen as a new but highly damaging disruption to the continuous efforts to reduce and eventually eradicate poverty globally. According to World Bank's report regrettably, 74 of the world's lowest-income countries are most impacted by the effects of the climate change and they are only responsible for one-tenth of the emissions of global greenhouse gases. In addition, it is not just the current poor nations that will be suffering as climate change is estimated to further push more than 100 million people into poverty by 2030 (Hallegatte et al. 2016). The utmost concerning issue in looking at the effects of climate change is that millions of people in the weakest position are facing inequal encounters with extreme weather events, food and water scarcity or contamination, unhygienic conditions, and health effects. People in the most affected regions are exposed due to the need for migration or forced displacement and are facing feeling of insecurity. They may lose their cultural connection and identity when they have to migrate or change their residency or occupation.

When we talk about the concept of *climate risk* it reflects both sociological and economic impacts and countries exposure to the various extreme weather events, such as floods, storms, droughts, heat waves, and the rise of sea level. The Germanwatch Institute first reported the results of their analysis on Global Climate Risk Index (GCRI) in 2020 during The UN Climate Change Conference COP 25 in Madrid. They have continued to update the report, and the latest GCRI 2021 report shows the impacts of extreme weather and the socioeconomic losses globally (Eckstein et al. 2021). The study analyzes and ranks the countries and regions and the level of impact. According to the data between 2000 and 2019, there have been 11,000 extreme weather events and almost 500,000 people has lost their lives. Also, the financial has been estimated to be as high as US$2.56 trillion. These events have created an increased activity to invest into adaptation with the estimated range of US$140 billion to US$300 billion per year by 2030 and will increase further amounting to US$280 billion to US$500 billion by 2050. Similarly, the IPPC reported that the cost of climate change by 2100 will be between US$ 54 trillion and US$ 69 trillion (IPCC Special Report 2018).

The most climate change-affected countries in 2019 were Mozambique, Zimbabwe, Malawi, and the Bahamas. During that same year one of the deadliest and most costly tropical storms, Tropical Cyclone Idai, hit the south-east coast of Africa, with extreme wind speeds of the order of 120 miles per hour and major torrential rain, flash floods, and landslides causing over 1000 casualties and effecting hundreds of thousands of people (Eckstein et al. 2021). Such impact and damage destroyed infrastructure, destroying total road networks, manufacturing facilities, and agricultural lands. In Malawi, the devastation affected one million people and left more than hundred thousand homeless and highly important agricultural sector badly damaged. Similarly, the Bahamas faced a category five Hurricane–Dorian. In this small island nation, the estimated damage was US$ 3.4 billion, and in one of the islands, the damage was on 45% of homes with more than 10,000 houses impacted. In the same year, Afghanistan, India, South Sudan, Niger, and Bolivia all encountered a series of extreme climate events, such as floods, landslides, heavy rainfall, droughts, and wildfires. These events have had significant impacts on these countries, causing widespread damage and posing challenges to their populations. All these countries except for Bahamas are, either developing countries or emerging low-income economies. Common to all is that they are either coastal or archipelagic.

According to the Global Climate Risk Index (GCRI) the most affected nations in 2019 were Puerto Rico (unincorporated US territory), Myanmar, Haiti, Philippines, Mozambique, and the Bahamas. The index also reports the long-term average values for extreme climate events during 2000–2019 period. According to the data, all the 2019 indexed countries except for Mozambique and Bahama are included with the addition of Pakistan and Bangladesh. These countries are continuously affected by climate change effects that dramatically disrupt their social and economic structures.

8.2.2 Island Nations

The most vulnerable island region is the Pacific Islands. There are more than 25,000 islands, and they are distributed over a region that covers more than 15% of the earth's surface. The population is only 2.3 million; however, the Islands have rich marine ecosystems and distinctly diverse natural inhabitants (Miloslavich et al. 2010). The impacts of climate change are first experienced in these islands. Firstly, the Pacific Islands are low-lying, with coral reefs and atolls. The highest point of this island is only about 5 m, and average elevation is about 2 m. With this elevation, the sea-level rise is of utmost risk as it has risen already in the past 30 years about 0.3 m. The estimated sea-level rise by 2050 is another 0.3–0.6 m (Birch and Wachter 2011). This creates a major risk for flooding, coastal erosion, and vulnerability to storm surges and upwelling.

In addition, the volatility of the extreme weather conditions has increased the frequency and the strength of the storms and the severity of the droughts. The acidification of the ocean is threatening the livelihood coming from fisheries and damages the coral reefs, which also are the very foundation of the islands and protects them. Also, freshwater supplies are threatened by these events and can destroy the local agriculture and heat stress is also increasing with high temperatures causing illness or even loss of life. The worst estimates are indicating that after 2050 the islands will become unstable and will create insupportable risks for human population (Campbell 2022).

Interestingly the annual number of tropical cyclones (named "hurricanes" in the eastern and central North Pacific, "typhoons" in the northwestern Pacific, and "cyclones" in the South Pacific) has been quite stable (82 $\pm$ 8 standard deviation) since 1981. It is estimated that this will continue to be the case in the long term as we'll (Schreck et al. 2014). However, the combination of the sea rise level and other climate change impacts with the tropical cyclones is creating expanded weather risks and effects.

Some of the climate change adaptation approaches include rainwater harvesting to prepare for droughts, erosion prevention, and flooding by using local plants introduced to the coastline. As the sea level enters the land the salt level increases, and therefore agriculture needs to use develop crops and plants that are more salt-tolerant. Economically the Pacific Islanders need to diversify their occupations and businesses. This includes for example expansion of their breeding, growing, and harvesting fish, shellfish, and selected aquatic plants. Emergency shelters need to be increasingly constructed and house floors and some of the selected sensitive infrastructure such as electric utilities and air-conditioning systems need to be raised higher than grounds level.

8.2.3 Coastal Cities and Metropolitan Areas

Although cities account for about 50% of world population, they only represent 1% of the total earth's surface area and their impact to global greenhouse emissions is as high as 75%. Cities with 10 million inhabitants or more are defined as megacities. Currently, there are 33 megacities around the world and the number and size are growing and they are facing major climate change risks, more specifically with storms, heat, flooding, and drought. It is also estimated that by 2050, 800 million people in 570 cities will face risks specifically in terms of the rise of sea level (Harper 2019). One issue with regard to the vulnerability and adaptation to climate change is that 80% of the megacities are located in developing countries. Interestingly, due to the high population growth rate, most of the expansion is in Africa and Asia and unfortunately some of these regions already experience challenging social issues in terms of poverty, unemployment, and inequality.

The majority of vulnerable megacities are in coastal areas, and it has been estimated that in terms of developing countries. Mumbai, India, is considered as the most climate-affected megacity. It is every year experiencing major floods, waterborne diseases, erosion, and landslides. Another highly vulnerable megacity is Dhaka in Bangladesh. It is situated only about a few meters above sea level with a heavy impact risk from storms and flooding.

Other cities in the developing countries with high climate change risk include Jakarta (Indonesia), Manila (Philippines), Calcutta (India), and Phnom Penh (Cambodia). In addition, Ho Chi Minh City in Vietnam that generates a quarter of the country's total GDP is highly vulnerable to especially sea-rise level together with the socioeconomically highly important city of Shanghai in China (Hoegh-Guldberg et al. 2019; Thomas et al. 2019).

The challenge for vulnerable megacities is that the climate change commands urgent and committed actions in terms of adaptation schemes and building socioeconomical resilience. Although the climate-related risks are clear and a major long-term risk to coastal regions, there is no trend in migration or relocation. Nor is there a clear sign of slowdown in population growth in these areas. Interestingly, due to climate change effects one of the most populated megacities, Jakarta (10 million people and 30 million more in metropolitan area), will lose its capital city status to Nusantara in the eastern Borneo by 2024. This is partially because of land sinking, caused by deep groundwater extraction due to seawater effect and other reasons including pollutants.

As these population hubs are also the highest emitters, if population growth were to slow down, the impact of climate change would be much reduced. Obviously, if all the positive global efforts for climate change mitigation are effective enough and according to plans, they can slow down the threats and risks felt by coastal cities and metropolitan areas.

When we look at a study by Nicholls et al. (2007), it describes the population exposure to coastal flooding in the 2070s, considering the climate change and socioeconomic change (Nicholls et al. 2007). The study ranks the top coastal areas exposed in 2007 and future (2070) population and exposed assets. Exposed population ranking includes 1. Kolkata (India), 2. Mumbai (India), 3. Dhaka (Bangladesh), 4. Guangzhou (China), and Ho Chi Minh City (Vietnam). It should be also noted that cities with high population exposure to sea levels also have the greatest exposure to *wind damage from tropical and extratropical cyclones*. The exposed population from this perspective ranged from 2.8 to 0.84 million. The research shows that the *exposure to coastal flooding* is high in large port cities with an estimated 40 million people or 10% of the total port city population. When it comes down to assets the list is different and includes 1. Miami (USA), 2. Guangzhou (China), 3. New York-Newark (USA), 4. Kolkata (India), and 5. Shanghai (China). The exposed assets in 2007, ranged from $420 billion to $32 billion. The study also listed the top 15 countries by population exposed, looking at both 2007 and the 2070s. The ranking in country level was China, India, Bangladesh, Vietnam, and the USA. From the asset exposure impact perspective, the list was somewhat different and the top five countries are China, USA, India, Japan, and Netherlands.

Various resident infrastructure and communities' programs have been initiated for climate change adaptation. These and other related programs are major but also necessary investments for the reduction of climate risks. The increasing cost of adaptation in developing countries is projected to reach $300 billion per year by 2030 (UN Environmental Program 2021). It presents a significant challenge for these countries. This cost is driven by various factors, including the impacts of climate change such as extreme weather events, sea-level rise, and changing ecosystems, which require investments in infrastructure, technology, and capacity building to build resilience and adapt to these changes.

Unfortunately, the current level of global adaptation finance falls short of the requirements for developing countries. In 2020, the total global adaptation finance amounted to just $46 billion, with a mere $28.6 billion allocated to support the adaptation efforts of developing nations (UN Environmental Program 2021). The observed funding gap highlights the urgent need for increased financial support to ensure that vulnerable countries can effectively adapt to the challenges posed by climate change. This leaves a substantial gap between the projected cost of adaptation and the available finance, posing a significant risk to vulnerable communities in developing countries.

The shortfall in adaptation finance has several consequences. It can hinder the implementation of adaptation measures, limit the capacity of developing countries to respond to climate impacts, and exacerbate inequality by disproportionately affecting marginalized communities who have limited resources to adapt. Insufficient adaptation finance can also result in increased debt burdens for developing countries as they may resort to borrowing to finance adaptation efforts, potentially leading to long-term economic challenges.

Closing this financial gap is of utmost importance in order to bolster the resilience of developing nations and enable them to effectively respond to the impacts of climate change. Adequate financial support is essential to enhance their capacity to adapt, implement adaptation measures, and build a sustainable future in the face of changing climatic conditions. By addressing the finance gap, we can empower these countries to strengthen their resilience, protect vulnerable communities, and achieve sustainable development goals in the midst of a rapidly changing climate. Increased financial resources are required from various sources, including public and private sectors, international financial institutions, and developed countries need to honor their commitments to climate finance. Innovative approaches are needed including climate risk insurance, public–private partnerships, and leveraging investments in climate technologies.

Despite contributing the least to global greenhouse gas emissions, developing countries bear a disproportionate burden when it comes to the impacts of climate change. These nations often lack the resources, infrastructure, and adaptive capacity needed to effectively respond and adapt to the changing climate conditions. As a result, they face heightened risks and vulnerabilities, including extreme weather events, sea-level rise, food and water scarcity, and ecosystem disruptions.

This disparity highlights the need for global cooperation and support to address climate change in a fair and equitable manner, ensuring that developing countries receive the necessary assistance and resources to mitigate and adapt to the challenges they face. They have already experienced a loss of 20% of GDP in the 55 most climate-vulnerable economies (OECD 2022). This is a stark reminder of the urgent need for increased climate adaptation finance, particularly in developing countries.

Bridging the gap between the projected cost of adaptation and the available finance is crucial in order to support developing countries in building resilience to climate change. International collaboration is indispensable in addressing the financial disparity faced by developing countries when dealing with the consequences of climate change, despite their minimal contribution to global greenhouse gas emissions.

By working together at a global level, it is possible to increase the flow of financial resources to these nations, enabling them to adapt to climate change impacts and protect their communities and ecosystems. This necessitates the active involvement and commitment of developed countries, international organizations, and financial institutions. Fostering collaboration and prioritizing the needs of developing countries, we can ensure the allocation of adequate financial resources to support their climate adaptation endeavors.

8.2.4 Socioeconomical Impact

The effects of climate change can have severe repercussions on livelihoods, especially in sectors like agriculture, forestry, and fisheries that rely heavily on climate-sensitive natural resources. Variations in temperature, rainfall patterns, and the occurrence of extreme weather events can significantly impact crop productivity, livestock health, and infrastructure integrity. Consequently, communities may experience income decline, food insecurity, and even displacement. These disruptions further emphasize the urgent need for comprehensive climate adaptation strategies to safeguard livelihoods and ensure the well-being of vulnerable populations.

Climate change poses a significant risk of incurring direct economic losses as a consequence of the destruction and impairment of infrastructure, property, and valuable assets caused by extreme weather events like hurricanes, floods, and wildfires. Such events can inflict substantial financial burdens on communities and economies, requiring extensive resources for rebuilding and recovery efforts. By exacerbating the frequency and intensity of these disasters, climate change underscores the importance of implementing effective mitigation and adaptation measures to minimize economic losses and ensure long-term sustainability.

These losses can affect local and national economies, disrupt supply chains, and impact businesses, leading to reduced economic growth and increased costs for governments, households, and the private sector. All industries will in a way, or another be impacted by climate change. The severity of it depends on the location and the value chain structure specific to the businesses and population serving it. The economic impact connects to the well-being of the localities and thus the social impact from climate change. Some industries do have significantly higher level of risk compared to the others and a different impact to social structures due to the location of the business operations.

The climate change impact to *traditional industry, retail and commercial services, tourism, insurance, and agriculture* are some of the most noticeable (USGCRP 2018). The *traditional industry* does experience both direct and indirect impacts. The industrial infrastructure and services including electrical power, sewage and water that are necessary for continuous operations can be impacted by extreme weather events. In addition, global supply chains can be highly sensitive to climate change. For example, ports that are a vital part of a maritime transportation system, are impacted by rising sea levels, increased precipitation, erosion, and storms. The energy sector, and more specifically, oil and gas industry are critical to industrial production and the overall supply chains and are both highly vulnerable and locally important. For example, extreme weather can impact beverage and food manufacturing and the necessary clean water supply from water treatment plants.

Retail industry is also highly dependent to supply chain and connects to the functionality and efficiency of the distribution of goods. This is especially important and a high risk in the coastal areas. Major sections of road and railroad infrastructure can be at risk in low-level parts of the transportation networks. Even in the case of small disruptions, the network effect can cause significant impacts to the entire business value chains.

The *insurance industry* is by its very nature a clear climate risk carrier due to the increasing number or frequency of insurance claims filed due to damage from extreme climate events. The other issue is the increased real estate values in areas that are at more risk, such as coastal properties. This is also connected to the increased population and rise of the middle class in developing countries. Major storms have historically been the most financially harmful events and insurance companies need to continuously improve their accuracy to predict the rate of recurrence and gravity of the incurred losses to the insured property. This knowledge needs to then transfer to the insurance premiums effectively.

Tourism is one of the highly impacted industries. The rapid global warming will impact the winter activity sector. In a short term, there are locations that will actually benefit from the additional precipitation that is connected to unchanged temperatures. However, in a long term as winters become shorter and the temperatures will gradually increase, the added precipitation will severely damage the winter snow conditions and the "cold weather" tourism will lose its ability to continue with the traditional business models. It has been estimated that by the year 2100 more than 60% of European ski reports will go out of business (New York Times 2014). As the water temperatures increase, tourism that is connected to coastal recreational and recreational fishing activities will suffer from the water quality degradation that comes from for example increased algae blooms causing toxicity. Also, island destinations and cruise harbor locations will be affected. Other impacts include forest fires and the reduced air quality in combination with deforestation and loss of biodiversity. This can detrimentally affect certain tourism destinations.

Agriculture and fishing are both connected to rural population and their commerce. As the temperatures increase, many crops can also grow faster. However, with faster growth speed the seeds will have less time to completely develop, and thus, the yield will go down. Some estimates show that the food prices will increase dramatically by 2050 (Falcon et al. 2022). The other effect with heat is drought and fires that will also reduce the ability to practice farming and will also reduce the yields. All this is concerning as an estimated 30% of the world's population works in agriculture.

Fishing industry is also facing adverse effects from climate change. Some species including salmon and trout are used to colder water temperatures and therefore will face habitat losses. The same issue also creates changes in predator–prey balance and interactions and will reduce the ability for some species to keep their traditional habitats. This causes fisheries the need to change their operating areas farther and thus will cause additional cost in time and money. The CO_2 levels in the atmosphere will also create ocean acidification and this impacts the oyster, clams, and shellfish business (Cinner et al. 2022).

The consequences of climate change on health are profound and encompass various aspects, including heightened morbidity and mortality rates attributable to heat stress, shifts in disease patterns, and heightened vulnerabilities to water and food-borne illnesses. Rising temperatures can lead to heat-related illnesses and heatwaves, posing risks to vulnerable populations, such as the elderly and individuals with pre-existing health conditions. Changes in climate can also influence the transmission patterns of infectious diseases, impacting the spread of vector-borne diseases like malaria or dengue fever. Furthermore, disruptions in water availability and quality, coupled with altered agricultural practices, can contribute to an increased risk of water- and food-borne diseases. Safeguarding public health in the face of climate change necessitates comprehensive strategies that prioritize adaptation, resilience, and the promotion of sustainable practices. Various health effects can put strain on healthcare systems, increase related costs, and reduce productivity, particularly in vulnerable populations with limited access to healthcare and other resources.

Climate change-induced *displacement and migration* pose complex challenges that transcend national borders and require comprehensive attention. The escalating risks of natural disasters, loss of livelihoods, and adverse environmental conditions can uproot communities, leading to both internal and cross-border movements. The consequences of displacement and migration can be far-reaching, affecting social structures, cultural identities, and economic stability. In receiving areas, there may be strains on social services and infrastructure as communities accommodate the influx of displaced individuals. Moreover, climate-related migration can exacerbate conflicts over limited resources and land, adding to existing sociopolitical tensions. Addressing these multifaceted challenges necessitates robust policies and collaborative efforts that prioritize the protection and well-being of affected populations, while also addressing the underlying causes of climate change and promoting sustainable development.

Climate change exacerbates existing *social inequalities and vulnerabilities*, disproportionately impacting marginalized and vulnerable populations. Low-income communities, indigenous peoples, women, and children are particularly affected by the adverse consequences of climate change. These groups often face multiple barriers, including limited access to resources, information, and adaptive capacity, which reduces their ability to cope with and recover from climate-related impacts. As a result, social inequities are amplified, deepening the disparities between different segments of society. It is crucial to address these disparities and prioritize the needs and rights of marginalized communities in climate change adaptation and mitigation strategies. This includes enhancing access to resources, empowering marginalized groups, promoting inclusive decision-making processes, and ensuring equitable distribution of benefits from climate action. By addressing social inequalities, we can foster resilience, justice, and sustainability in the face of climate change.

The *cost of adaptation* to climate change impacts can pose a significant socioeconomic burden, particularly for developing countries and vulnerable communities.

The implementation of adaptation measures to address the impacts of climate change necessitates adequate financial resources, technical expertise, and institutional support. However, many regions face challenges in accessing and mobilizing these necessary resources. Limited financial capacity, technological gaps, and weak institutional frameworks can hinder the effective implementation of climate resilience initiatives. This creates a need for increased support and collaboration at local, national, and international levels to build the necessary capacity, strengthen institutions, and secure funding for adaptation projects. By addressing these barriers and providing the required resources and support, communities and regions can enhance their resilience to climate change and minimize the adverse impacts on their social, economic, and environmental systems. Lastly, *security risks* have implications, including conflicts over resources such as water and land, competition for migration routes and safe havens, and potential displacement-driven social and political instability. The socioeconomic impacts of these risks include disruptions to trade, investments, and tourism, and can further exacerbate existing vulnerabilities and inequalities.

Climate change has wide-ranging socioeconomic impacts that can affect various aspects of societies and economies, with vulnerable populations being disproportionately affected. Addressing these impacts requires comprehensive strategies that include adaptation measures, social protection mechanisms, equitable policies, and international cooperation to mitigate and manage the socioeconomic consequences of climate change.

8.2.5 Mitigation Efforts in Cities and Municipalities

Sustainable city is a concept with a purpose of addressing social, environmental, and economic design of cities and municipalities. It encompasses various aspects of sustainability ranging from waste management and water conservation to green buildings and public green spaces. Although the focus is seemingly on environmental aspects and eco-city approaches, sustainably city is a more systemic approach to sustainability and aims at improving the quality of urban citizen's life.

The eco-city concept is more concentrated on urban planning and management and was introduced as early as 1975 by California (Berkeley)-based non-profit Urban Ecology. From its beginning, eco-city included urban design concepts that integrated urban planning, ecological systems, and citizen participation. Although in principle both concepts are very similar, in practice eco-city can be considered as any urban sustainability initiative (Bibri and Bibri 2020).

Sustainable cities and eco-cities are not specifically designed to mitigate climate change but as such they are an important part of the equation. They develop solutions to reduce costs and to create an urban sustainability culture that facilitates the advancement of carbon neutrality. Many aspects of redesigning cities clearly connect to carbon reduction including infrastructure investments, building bike lanes and more walkable communities, or green transportation and greener energy, power, and utilities. The investments are long term and by time enable sustainable carbon zero approaches to cities and municipalities.

An important part of the impetus and encouragement of building sustainable cities is rankings and competitions. One example is *Arcadis Sustainable Cities Index.* It is based on 51 metrics using 26 different indicator themes overarching the balance of the three pillars of planet, people, and profit. The index gives an indicative ranking of top 50 sustainable cities globally, and it was founded in 2015 (Arcadis 2023). The index shows overall rankings, and rankings for all three pillars of sustainability (people-social, planet-environmental, profit-economic). The ranking top five cities in 2022 were 1. Oslo, 2. Stockholm, 3. Tokyo, 4. Copenhagen, and 5. Berlin.

Corporate Knights is a sustainable-economy media and research company based in Canada and established in 2002 by Toby A. Heaps, Paul Fengler, and Peter Diplaros. They have a research division that makes global sustainability rankings and research reports. One of their indexes is *Sustainable Cities Index.* It is focused on environmental performance and climate resilience in global cities. The methodology is highly quantitative, using different indicators to evaluate the global cities sustainability status. The main focus is currently on environmental issues, but the index is looking at adding also social dimension to the evaluation. The data collected and used is mainly based on physical measurements or indicators such as water and air quality, waste management, walkability and local greenhouse emissions and resiliency for climate change. Also, one indicator is policy-related and focused on the city's commitment to making investments to sustainable transportation, reduction of greenhouse gases, and renewable energy in power and utilities infrastructure and purchases. As an important part of the evaluation process stakeholder feedback is actively solicited throughout the project. The 2022 reports show the following cities to be in the top five groups: 1. Stockholm, 2. Oslo, 3. Copenhagen, 4. Lahti, and 5. London. Compared to Arcadis index, there are similarities in the ranking; however, the difference is due to the social and economic bias in the Arcadis index methodology.

Other indices worth mentioning include *The Green City Index* that was launched in 2008 as a collaborative effort between the Economist Intelligence Unit (EIU) and Siemens. It evaluates the environmental sustainability of 30 European cities. The methodology uses 30 indicators including CO_2 emissions, energy use, building sustainability, land use, sustainability of transportation, water and waste management, air quality, and environmental governance. It has evolved to also include cities across Asia, Africa, and the Americas.

CITYkeys Indices is a European smart city indicator framework that has been developed recently (CITYkeys 2017). It was made as part of a research project with an objective to create a city performance framework. The approach to creating Key Performance Indicators (KPIs) for the CITYkeys performance measurement system was to analyze different needs for development and cities' needs, using existing indicators and gaps. As this indicator is focused on smart cities, it includes topics connected to digitalization and its role in sustainable cities. The broad range of topics covered includes energy systems, reduction of greenhouse gas emissions, sustainable transportation, digital infrastructure and e-services, efficient resource management, citizen participation, economic competitiveness, environmental conservation, improving quality of life, and promoting research and knowledge creation. These topics collectively address various aspects of addressing climate change challenges and creating a sustainable and resilient future. By focusing on these areas, stakeholders can work towards developing innovative solutions, policies, and strategies that contribute to mitigating climate change and fostering sustainable development.

A high-level incentive for sustainable city or environmental city development is the *European Commission's Green Capital Award*, which is purposed to give the highest recognition and reward to local efforts in improving the environment, and local economy and in general the quality of life (European Green Capital Award 2023). This is an annual award, and it is given each year to a leading city that shows excellent progress and continuous commitment to further environmental improvement. The evaluation process is based on seven environmental indicators: air quality, water quality and efficiency, biodiversity, green areas, sustainable land use, waste and circular economy, and noise. In addition, climate change is also specifically addressed looking at mitigation, energy performance, and adaptation. The award was bestowed to Stockholm, Sweden in 2010. The past 5 years the winners have been Essen, Germany (2017), Nijmegen, Netherlands (2018), Oslo, Norway (2019),

Lisbon, Portugal (2020), Lahti, Finland (2021), Grenoble, France (2022), and Tallinn, Estonia (2023).

Cultural and mindset change play a crucial role in addressing the social aspects of climate change and promoting effective mitigation efforts in cities and municipalities. Recognizing that climate change is not solely an environmental issue, but also a social one, is essential for understanding its impacts on marginalized and vulnerable communities. By fostering a cultural shift and promoting a mindset of environmental stewardship, inclusivity, and social equity, societies can work towards addressing climate change in a more holistic and equitable manner. This involves raising awareness, promoting education, and engaging communities in decision-making processes to ensure that climate actions consider the diverse needs and perspectives of all individuals and groups. It also involves fostering a sense of responsibility, empathy, and collective action to create a sustainable and resilient future for everyone.

Cultural and mindset change is needed in order to address these social aspects of climate change. The key is in promoting greater awareness and understanding of the ways in which climate change affects different communities. For example, breaking down stereotypes and promoting greater empathy and support for those who are in most need and affected by climate change, including indigenous peoples, low-income communities. Greater equity is needed in addressing climate change, by recognizing the social and economic factors that contribute to climate change. This can help promote policies and initiatives that prioritize the needs and perspectives of marginalized communities.

In addition, promoting greater collaboration and cooperation within communities between different groups and individuals is needed. Recognizing the interconnectedness of social and environmental issues can help foster greater dialogue and understanding between governments, civil society organizations, and other stakeholders. Changing cultural and mindset norms requires addressing deep-seated beliefs and values that have been passed down through generations. This can involve education, awareness-raising campaigns, and community engagement efforts. It may also require policy changes that encourage sustainable behaviors and lifestyles.

Overall, cultural and mindset change is a critical component of addressing climate change. If a culture values sustainability and environmental responsibility, individuals within that culture may be more motivated to take actions to mitigate climate change. This is critical in building sustainable cities and communities.

References

Arcadis (2023). https://connect.arcadis.com/Sustainable-Cities-Index

Bibri SE, Bibri SE (2020) Advances in eco-city planning and development: emerging practices and strategies for integrating the goals of sustainability. Adv Lead Paradigms Urbanism Amalgamation Compact Cities Eco Cities Data Driven Smart Cities

Birch EL, Wachter SM (eds) (2011) Global urbanization. University of Pennsylvania Press

Campbell JR (2022) From the frying pan into the fire? Climate change, urbanization and (in)security in pacific island countries and territories. Peace Rev 34(1):11–21. https://doi.org/10.1080/10402659.2022.2023425

Cinner JE, Caldwell IR, Thiault L, Ben J, Blanchard JL, Coll M, Diedrich A, Eddy TD, Everett JD, Folberth C, Gascuel D, Pollnac R (2022) Potential impacts of climate change on agriculture and fisheries production in 72 tropical coastal communities. Nat Commun 13(1):3530

CITYkeys (2017). https://publications.vtt.fi/julkaisut/muut/2017/CITYkeys_smart_city_performance_measurement_system.pdf

Eckstein D, Kunzel V, Schafer L (2021) Global climate risk index 2021. German watch. https://www.germanwatch.org/en/cri

European Green Capital Award (2023). https://environment.ec.europa.eu/topics/urban-environment/european-green-capital-award/winning-cities_en

Falcon WP, Naylor RL, Shankar ND (2022) Rethinking global food demand for 2050. Population Dev Rev 48(4):921–57

Harper S (2019) The convergence of population ageing with climate change. J Population Ageing 12:401–403

Hoegh-Guldberg O, Jacob D, Taylor M, Guillén Bolaños T, Bindi M, Brown S, Camilloni IA, Diedhiou A, Djalante R, Ebi K, Engelbrecht F, Zhou G (2019) The human imperative of stabilizing global climate change at 1.5 °C. Science 365(6459):eaaw6974

IPCC Special Report (2018). Retrieved from https://www.ipcc.ch/site/assets/uploads/sites/2/2019/06/SR15_Full_Report_High_Res.pdf. 18 Mar 2022

New York Times (2014). https://www.nytimes.com/2014/02/08/opinion/sunday/the-end-of-snow.html

Nicholls et al (2007) OECD, Paris

OECD (2022) Aggregate trends of climate finance provided and mobilised by developed countries in 2013–2020. https://www.oecd.org/climate-change/finance-usd-100-billion-goal

Schreck CJ, Knapp KR, Kossin JP (2014) The impact of best track discrepancies on global tropical cyclone climatologies using IBTrACS. Mon Wea Rev 142:3881–3899. https://doi.org/10.1175/MWR-D-14-00021.1

Thomas K, Hardy RD, Lazrus H, Mendez M, Orlove B, Rivera-Collazo I, Roberts JT, Rockman M, Warner BP, Winthrop R (2019) Explaining differential vulnerability to climate change: a social science review. Wiley Interdisc Rev Clim Change 10(2):e565

United Nations Environment Programme (2021) Adaptation gap report 2021: the gathering storm—adapting to climate change in a post-pandemic world. Nairobi. https://www.unep.org/adaptation-gap-report-2021

USGCRP (2018) Impacts, risks, and adaptation in the United States: fourth national climate assessment, vol 2. (Reidmiller DR, Avery CW, Easterling DR, Kunkel KE, Lewis KLM, Maycock TK, Stewart BC)

Wei T, Wu J, Chen S (2021) Keeping track of greenhouse gas emission reduction progress and targets in 167 cities worldwide. Front Sustain Cities 3

Chapter 9
Yellow Brick Road: Roadmap for 2050

Abstract In the concluding chapter of this book, we present a comprehensive roadmap for addressing climate change and achieving carbon neutrality. The climate actions typically involve a two-step process. The first step emphasizes adopting sustainable practices, improving energy efficiency, transitioning to renewable energy sources, and implementing emissions reduction measures in various sectors. The second step involves offsetting remaining emissions through carbon offset projects or using Carbon Capture and Storage (CCS) technologies. These projects focus on activities that remove or reduce greenhouse gases from the atmosphere, such as afforestation, reforestation, soil carbon sequestration, and renewable energy initiatives. Achieving carbon neutrality comes with significant challenges and requires substantial investments. Funding is critical to support mitigation efforts and can be obtained through various means, including investments, bonds, grants, and other financial mechanisms. Governments, private companies, philanthropic organizations, and individuals can contribute to these funding initiatives, directing resources towards activities aimed at addressing climate change and promoting sustainability. The roadmap to carbon neutrality is complex and demanding. In this book, the proposed action plan involves two distinctive and parallel approaches: culture and mindset change and technological solutions. Raising awareness, fostering sustainable values, empowering community engagement, and advocating for policy changes are vital for change. On the technological front, investments in renewable energy, energy efficiency, CCS technologies, hydrogen technology, and afforestation and reforestation are paramount. The final chapter emphasizes the urgency and collective responsibility to address the climate crisis, paving the way towards a sustainable and carbon–neutral future.

Keywords Roadmap · Carbon neutrality · Off-setting · Afforestation and reforestation · Culture and mindset change · Technological solutions

Carbon neutrality is a principle that aims to achieve a balance between the amount of GHG emissions generated and the amount of GHG removed or offset from the atmosphere. It involves taking actions to minimize GHG emissions through various

M. Hakovirta, *Carbon Neutrality*, Springer Climate,
https://doi.org/10.1007/978-3-031-45202-4_9

means, such as implementing energy-efficient practices, transitioning to renewable energy sources, and adopting sustainable practices across different sectors. Additionally, carbon neutrality often involves supporting projects or initiatives that remove or offset an equivalent amount of GHG emissions, such as investing in reforestation projects or supporting renewable energy projects. By striving for carbon neutrality, organizations, communities, and even countries aim to reduce their overall carbon footprint and contribute to mitigating climate change.

Achieving carbon neutrality typically involves a two-step process: *reducing emissions* and *offsetting remaining emissions*. The first step involves *adopting sustainable practices*, improving energy efficiency, transitioning to renewable energy sources, and implementing measures to reduce emissions from various sectors, including energy, transportation, agriculture, and industry. This can include investing in renewable energy infrastructure, promoting energy conservation and efficiency, optimizing transportation and logistics, and implementing emissions reduction technologies and practices.

The second step in achieving carbon neutrality is *offsetting remaining emissions* through measures such as investing in carbon offset projects or utilizing CCS technologies. Carbon offset projects can involve activities that remove or reduce GHGs from the atmosphere, such as afforestation and reforestation, soil carbon sequestration, and renewable energy projects.

Carbon neutrality is often seen as a transitional step towards the broader goal of achieving carbon zero or a *carbon negative state*. Carbon zero refers to a state where there is no net release of CO_2 into the atmosphere, achieved by balancing emissions with equivalent removal or offset, while carbon negative refers to a state where more GHGs are removed from the atmosphere than emitted.

There are several motivations for entities, including companies, organizations, and countries, to pursue carbon neutrality. These can include environmental sustainability goals, addressing climate change risks and impacts, economic incentives, and new business opportunities, complying with regulatory requirements, meeting stakeholder expectations, and enhancing reputation and brand value.

However, achieving carbon neutrality comes with major challenges and need for investments. These can include accurately measuring and reporting emissions, verifying the effectiveness and credibility of offset projects, ensuring the integrity of climate technologies, and addressing issues of equity and fairness in the distribution of costs and benefits.

In recent years, there has been a growing recognition worldwide of the importance of carbon neutrality as a key strategy to address and mitigate the impacts of climate change. The urgency to reduce GHG emissions and limit global warming has led to emphasis on achieving carbon neutrality as a crucial milestone in the transition to a more sustainable and climate-resilient future. Governments, organizations, and individuals are acknowledging the need to take concrete actions to reduce their carbon footprint and adopt practices that result in no net increase in atmospheric GHG concentrations. This involves implementing measures to minimize emissions, promoting renewable energy sources, improving energy efficiency, and supporting projects that remove or offset emissions. The pursuit of carbon neutrality reflects a collective global effort to combat climate change and create a more sustainable planet for future generations. Many governments, organizations, and companies have set targets to achieve carbon neutrality by specific dates, and various voluntary and regulatory frameworks have been developed to guide and incentivize such efforts. Nevertheless, achieving carbon neutrality requires concerted and sustained efforts across multiple sectors and stakeholders.

Funding carbon neutrality is critical and involves providing financial resources to support mitigation efforts. This can be done through various means, such as investments, bonds, grants, and other financial mechanisms, to support emission reduction initiatives and carbon offset projects. Funding for carbon neutrality can come from various sources, including governments, private companies, philanthropic organizations, and individuals. These funds can be allocated towards a diverse array of activities aimed at addressing climate change and promoting sustainability. They can be used to support initiatives such as the development and implementation of renewable energy projects, which generate clean and sustainable sources of power. Additionally, funds can be directed towards energy efficiency initiatives that aim to reduce energy consumption and minimize waste. Another area where funds can make a significant impact is in supporting afforestation and reforestation efforts, which contribute to carbon sequestration and biodiversity conservation.

Energy is in the heart of climate change and therefore investments in for example renewable energy infrastructure, solar or wind power projects, will increase the availability of clean and low-carbon energy and reduce the reliance on fossil fuels lower GHG emissions. Energy efficiency initiatives play a crucial role in mitigating climate change by promoting the efficient use of energy across various sectors. By implementing energy-saving measures and technologies, such as improving insulation, upgrading lighting systems, and optimizing industrial processes, energy consumption can be significantly reduced. This, in turn, leads to a decrease in greenhouse gas emissions in terms of energy production and usage.

CCS technologies capture CO_2 emissions from industrial processes or power plants and store them underground, preventing them from entering the atmosphere. Their role may be critical in reducing emissions from hard-to-decarbonize sectors. However, significant funding is required to support further research, development, and deployment of these technologies. This funding is necessary to drive innovation, improve existing technologies, and accelerate their adoption on a larger scale.

Hydrogen technology is a recent entrant to the climate technology portfolio, and it also has the potential to play a significant role in mitigation efforts. Hydrogen is a clean-burning fuel that produces only water as a by-product, making it a fuel with zero emission. Hydrogen can also be used as an energy source in fuel cells, which convert hydrogen into electricity with water and heat as by-products. Fuel cells are highly efficient and can be used by themselves or as an additional power for a wide range of applications, from cars and trucks to homes and businesses. Hydrogen can be generated using various methods, including the electrolysis of water, the reforming of natural gas, and biomass gasification. Since hydrogen can be made using fossil fuels it is crucial to focus on using hydrogen from renewable sources. By doing so, we can ensure that hydrogen energy is a clean and sustainable solution to climate change.

Afforestation and reforestation efforts involve planting new trees or restoring degraded forests, which can act as carbon sinks by absorbing and storing CO_2 from the atmosphere. They also have a significant positive impact on biodiversity as well as climate change mitigation. Forests are home to a large proportion of the world's terrestrial biodiversity, providing habitat for countless species of plants, animals, and microorganisms. When forests are destroyed, species lose their habitats and may go extinct. Deforestation can also fragment habitats, leading to further loss of biodiversity. Additionally, by promoting the regeneration of native plant species, it can help to restore biodiversity and support ecosystem processes such as pollination, seed dispersal, and nutrient cycling. Afforestation and reforestation need to be carried out in a sustainable and socially responsible manner to avoid negative impacts on biodiversity. For example, monoculture plantations that focus on a single species or crop can have negative impacts on biodiversity, as they may not provide adequate habitat for a wide range of species. Therefore, it is important to consider the ecological and social impacts and to engage local communities in the planning and implementation of these projects to ensure that they are sustainable and socially responsible. In this context also, soil carbon sequestration projects aim to improve soil health and increase carbon storage in soils through practices such as conservation agriculture, agroforestry, and wetland restoration.

It is important to ensure that funding for carbon neutrality is used effectively and transparently, with clear measurement and reporting of the impact of funded initiatives on emissions reductions or removals. Verification of carbon offset projects and climate technologies' integrity is crucial to ensure that the funded projects achieve their intended goals and contribute to real emissions reductions. Funding carbon neutrality plays a critical role in accelerating the transition to a more sustainable and low-carbon economy. Financial resources are needed to support emission reduction initiatives and carbon offset projects, which can help mitigate climate change risks and impacts, contribute to global climate goals, and support the development of a sustainable and resilient future.

9.1 Roadmap to Carbon Neutrality

The road to carbon zero is long and challenging. This transition requires significant investments in research and development, infrastructure, and policy changes. It also requires a fundamental shift in our attitudes and behaviors towards energy consumption and waste reduction. Decarbonizing certain industries, such as aviation and heavy industry, poses significant challenges due to the limited availability of viable alternatives to fossil fuels. This will require innovative solutions and new technologies to address these challenges. The road to carbon zero is not impossible and many countries and companies have already made significant progress. As the climate technology landscape continues to advance and the costs of renewable energy decrease, it will become increasingly feasible to achieve carbon zero. It will require collective effort and political will, but the benefits of a sustainable and carbon-free future are worth the effort.

Based on the various aspects, this book has discussed and reviewed, a *summary of a proposed action plan* with steps to address the climate crisis can be summarized with two distinctive categories, *culture and mindset* and *technological solutions:*

9.1.1 Culture and Mindset Change

Raise Awareness and Education: Increase public awareness about the urgency of the climate crisis and the importance of collective action. Implement educational programs to inform individuals and communities about the impacts of climate change and the significance of sustainable practices. Encourage individuals to embrace environmentally conscious behavior including reducing energy consumption, adopting sustainable transportation options, and embracing eco-friendly lifestyles.

Promote Social Equity: Address social inequalities and vulnerabilities that are increased by climate change. Ensure that climate action initiatives prioritize the needs of marginalized and vulnerable populations and promote inclusive participation in decision-making processes.

Invest: Redirecting financial resources towards climate-friendly investments is a key *accelerant to climate change mitigation* and also for building adaptation pathways. This can involve divesting financial resources from fossil fuels and redirecting investments towards renewable energy, energy efficiency measures, and sustainable practices across various sectors. It can also involve supporting green finance initiatives, impact investing, and sustainable business practices.

Work on politics and policy: This is also an *accelerant* and needs specific attention. Advocacy and engagement in politics and policymaking are crucial to drive systemic change. This can involve supporting policies and regulations that promote renewable energy, carbon pricing, and emissions reduction targets, as well as advocating for climate action at all levels of government.

Create urgency: Mobilizing public support and building movements for climate action can create the necessary urgency and momentum to also *accelerate* and drive change. This can involve grassroots demonstrations, and community engagement to demand climate action from governments, businesses, and other stakeholders.

Innovate!: Encouraging and supporting innovation in climate technologies, sustainable practices, and climate solutions can expedite the shift towards a low-carbon economy. This can involve funding research and development, promoting collaboration between academia, industry, and government, and fostering a culture of innovation and entrepreneurship.

9.1.2 Technological Solutions

Electrify transportation: Transitioning to electric vehicles (EVs) and investing in charging infrastructure can have a substantial impact on reducing GHG emissions from the transportation sector, which is a significant contributor to climate change. This can be achieved through incentives for EV adoption, investment in EV charging stations, and also in general by promotion of public transportation, walking, and biking.

Decarbonize the grid: The transition to renewable energy sources such as wind, solar, and hydropower is essential to decarbonize the electricity grid and effectively reduce GHG emissions. This can be done through policies and incentives that promote renewable energy development, grid modernization, and energy storage technologies.

Resolve food: Adopting sustainable agriculture practices, reducing food waste, and promoting plant-based diets can help reduce GHG emissions from the food system. This can involve supporting regenerative agriculture, promoting local and sustainable food production, and raising awareness about the environmental impact of food choices.

Protect nature: Conserving and restoring natural ecosystems such as forests, wetlands, and oceans can sequester carbon and help mitigate climate change. This can involve protecting and restoring biodiversity-rich areas, implementing sustainable land management practices, and supporting initiatives to reduce deforestation and forest degradation.

Clean-up industry: Implementing clean technologies and practices in industrial sectors such as manufacturing, construction, and heavy industry can reduce emissions. This can involve adopting low-carbon production methods, improving energy efficiency, and investing in technologies such as hydrogen solutions, and small-scale nuclear power in high-emitting industries.

Remove carbon: The development and implementation of CCS technologies play a crucial role in reducing GHG concentrations by removing and storing CO_2 from the atmosphere. This requires research, development, and deployment of innovative carbon removal technologies and practices (Fig. 9.1).

Implementing these steps and actions collectively and comprehensively will mitigate the climate crisis and create a more sustainable and resilient future for our planet. It requires coordinated efforts from governments, businesses, communities, and individuals to transition to a low-carbon economy and address the urgent challenges posed by climate change. More specifically, the components in the roadmap: *politics and policy, urgency, innovations and investments* are necessary *accelerants* that refer to actions or strategies that can expedite or speed up the process of addressing the climate crisis.

Setting clear net zero goals is a critical step in addressing the climate crisis. Some key considerations for setting effective net zero goals include covering all scopes (1,

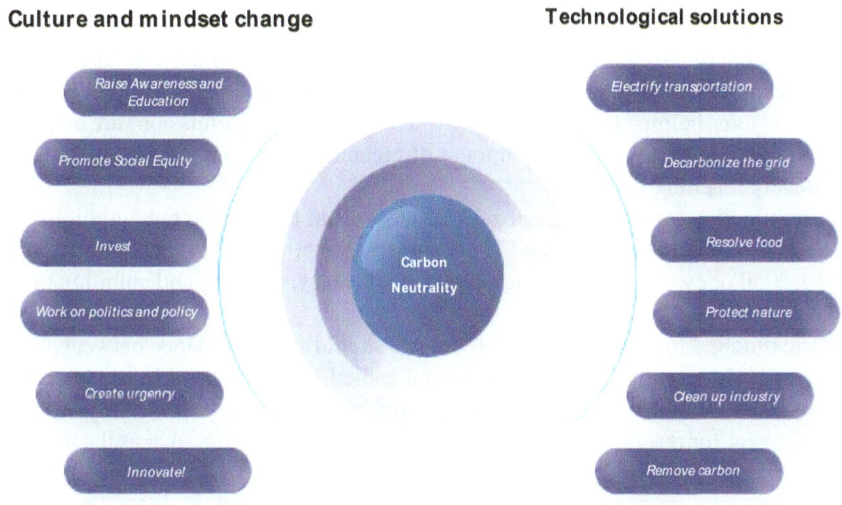

Fig. 9.1 Components of carbon neutrality roadmap

2, 3): net zero goals should encompass all three scopes of greenhouse gas emissions as defined by the Greenhouse Gas Protocol. Scope 1 includes direct emissions from owned or controlled sources, Scope 2 includes indirect emissions from purchased electricity, heat, or steam, and Scope 3 includes indirect emissions from value chain activities, such as those from suppliers, customers, and transportation. By addressing emissions across all scopes, organizations can comprehensively tackle their carbon footprint and work towards achieving net zero emissions.

A comprehensive net zero goal should address emissions from all scopes to ensure a holistic approach to reducing greenhouse gas emissions. Net zero goals should encompass all major greenhouse gases, including CO_2, methane (CH_4), nitrous oxide (N_2O), and fluorinated gases (F-gases). These gases have different global warming potentials and contribute to climate change to varying degrees. By addressing all of these gases collectively, we can effectively mitigate the impact of greenhouse gas emissions and work towards achieving meaningful climate action. This comprehensive approach ensures that all significant contributors to global warming are targeted, maximizing the potential for reducing greenhouse gas concentrations in the atmosphere and limiting the extent of climate change.

A true commitment and accountability to using removal is critical and therefore net zero goals should incorporate the use of carbon removal technologies or nature-based solutions that can remove or offset greenhouse gas emissions from the atmosphere. Practices such as reforestation, afforestation, soil carbon sequestration, and direct air capture play a crucial role in achieving negative emissions and effectively balancing remaining emissions to reach a net zero state. These approaches involve activities that remove CO_2 from the atmosphere or enhance carbon storage in natural and managed ecosystems. Reforestation and afforestation involve planting trees in areas where they were previously absent, while soil carbon sequestration focuses on improving soil health and increasing carbon storage in agricultural lands. Direct air capture technologies aim to capture CO_2 directly from the air and store or utilize it. By employing these practices, we can actively remove or offset greenhouse gas emissions, helping to achieve a state where any remaining emissions are balanced by the removal of an equivalent amount of greenhouse gases from the atmosphere, ultimately reaching net zero.

The timeline for achieving net zero targets should be realistic but ambitious and include a step-by-step roadmap. This can be set as a specific year, such as 2030, 2040, 2050 or even earlier, depending on the relevancy, urgency, and ambition of the goals.

The timeline for achieving net zero targets should strike a balance between being realistic and ambitious. It should be accompanied by a step-by-step roadmap that outlines the necessary actions and milestones to be achieved along the way. The specific year for reaching net zero can vary depending on factors such as the urgency of the climate crisis and the level of ambition set by each country or organization. Common target years include 2030, 2040, 2050, or even earlier, depending on the context. The chosen timeline should consider the latest advancements in technology and scientific knowledge, as well as the imperative to limit global temperature rise to well below 2 °C above preindustrial levels, as outlined in the Paris Agreement. By setting a clear and time-bound target, we can focus our efforts and resources towards a sustainable and low-carbon future.

In summary, establishing clear net zero goals that encompass all scopes and greenhouse gases, incorporating removal strategies, and adhering to an ambitious timeline is essential for impactful climate action. The roadmap towards achieving carbon zero must be a comprehensive and interconnected system that requires collaboration among organizations, governments, and communities. Each actor must assume responsibility and play a significant role in mitigating the impacts of climate change. By working together, we can create a sustainable and low-carbon future for generations to come.

Index

M. Hakovirta, *Carbon Neutrality*, Springer Climate,
https://doi.org/10.1007/978-3-031-45202-4

The manufacturer's authorised representative in the EU is Springer Nature Customer Service Centre GmbH, Europaplatz 3, 69115 Heidelberg, Germany. If you have any concerns regarding our products, please contact ProductSafety@springernature.com

Printed and bound by CPI Group (UK) Ltd, Croydon, CR0 4YY
28/09/2026
02224081-0019